New Trends in Fluid and Solid Models

Proceedings of the International Conference
in Honour of Brian Straughan

New Trends in Fluid and Solid Models

Proceedings of the International Conference in Honour of Brian Straughan

Vietri sul Mare (SA), Italy 28 February – 1 March 2008

editors

Michele Ciarletta
University of Salerno, Italy

Mauro Fabrizio
University of Bologna, Italy

Angelo Morro
University of Genoa, Italy

Salvatore Rionero
University of Naples "Federico II", Italy

World Scientific

NEW JERSEY • LONDON • SINGAPORE • BEIJING • SHANGHAI • HONG KONG • TAIPEI • CHENNAI

Published by

World Scientific Publishing Co. Pte. Ltd.
5 Toh Tuck Link, Singapore 596224
USA office: 27 Warren Street, Suite 401-402, Hackensack, NJ 07601
UK office: 57 Shelton Street, Covent Garden, London WC2H 9HE

British Library Cataloguing-in-Publication Data
A catalogue record for this book is available from the British Library.

NEW TRENDS IN FLUID AND SOLID MODELS
Proceedings of the International Conference in Honour of Brian Straughan

ISBN-13 978-981-4293-21-1
ISBN-10 981-4293-21-0

Printed in Singapore by World Scientific Printers

PREFACE

Dedicated to Brian Straughan

The International Conference "New Trends in Fluid and Solid Models 2007" has been dedicated to Prof. Brian Straughan in the occasion of his 60th birthday. For the sake of simplicity and concreteness we confine ourselves to mentioning the main points of his scientific and academic career. More informations can be found in his web site address.

Born in County Durham, England, 18th October, 1947.

Degrees

- B.Sc., Mathematics, 1970, University of Newcastle upon Tyne.
- M.Sc., Fluid Mechanics and Mathematical Techniques, University of Newcastle upon Tyne, 1971.
- Ph.D. in Mathematics, Heriot-Watt University, 1974.

Career

- Oct. 1966 – Sept. 1967: Chemist with Procter and Gamble Ltd., Newcastle upon Tyne.
- Oct. 1971 – Sept. 1972: Research Assistant, Department of Electrical Engineering, Liverpool University.
- Nov. 1974 – Sept. 1975: SERC Research Associate, Mathematics, Heriot-Watt University.
- Oct. 1975 – Sept. 1978: University Research Fellow, Mathematics, Glasgow University.
- Oct. 1978 – Sept. 1986: Lecturer, Glasgow University.
- Oct. 1986 – Sept. 1987: Senior Lecturer, Mathematics, Glasgow University.
- Oct. 1987 – Nov. 1991: Titular Professor, Mathematics, Glasgow University.
- December 1991 – August 2000: Simson Professor of Mathematics in the University of Glasgow.

- Currently, Professor of Numerical Analysis in the University of Durham, since Sept. 2000.

Memberships

- Elected a Fellow of the Royal Society of Edinburgh, March 1986.
- Elected a Foreign Member of the Accademia di Scienze, Fisiche e Matematiche di Napoli, 3rd June, 1995.

Web site address

http://maths.dur.ac.uk/ dma0bs/

Personal details

Married Carole Spark, April 1, 1972.
Two daughters, Luan (1973), Kirsten (1980).
Two grandsons, Cole Philip (2005), Caleb Lawrie (2009).

His scientific activity is concerned with partial differential equations, scientific computing and their applications to problems in the real world. He is author of more than 150 scientific papers concerned with

(1) Linear and non linear stability of dynamical systems;
(2) Energy methods;
(3) Numerical/Computational problems (in particular numerical solutions of spectral problems);
(4) Convection;
(5) Fluid/Porous flow in bounded and unbounded domains;
(6) Linear and non linear elasticity, viscoelastcity and wave propagation;
(7) Math. Biology/Medicine.

In the field of energy stability, he is one of the founders of the modern guide lines. He is the author of six research books, among which

(1) Stability and wave motion in porous media (Vol. 165, *Appl. Math. Sci. Ser.*, Springer-Verlag, 2008)
(2) The energy method, stability and non linear convection (Vol. 91, *Appl. Sci. Ser.*, Springer-Verlag, second edition, 2004)

Prof. Brian Straughan

CONTENTS

Well-Posedness for a Ginzburg-Landau Model in Superfluidity

V. Berti[*], *M. Fabrizio*[†]

Dipartimento di Matematica,
Università di Bologna,
40126 Bologna, Italy.

1 Introduction

Superfluidity was discovered in 1938 by Kapitza, Allen and Misener, who observed that at a temperature close to the absolute zero, an isotope of helium, ^{4}He was able to flow without viscosity in narrow capillaries. More precisely, immediately below its boiling temperature $\theta = 4.21K$, ^{4}He behaves like an ordinary fluid with small viscosity. However it exhibits a superfluid behavior, when the temperature decreases under the critical value $\theta_\lambda = 2.17K$.

The traditional theories of superfluidity consider a superfluid as a mixture of two fluids, with the density ρ and the velocity $\mathbf{v}$ given by

$$\rho = \rho_n + \rho_s \ , \qquad \rho\mathbf{v} = \rho_n\mathbf{v}_n + \rho_s\mathbf{v}_s.$$

Here ρ_n, ρ_s, $\mathbf{v}_n$, $\mathbf{v}_s$ are called respectively normal and superfluid density and velocity. The two fluid model was suggested by Tisza in order to explain experiments showing the twofold behavior of liquid helium: on the one hand it is able to flow frictionless through narrow channels, on the other, a body moving in ^{4}He shows the existence of a viscous drag.

According to Landau's point of view, the two-fluid model is just a convenient way of saying that each particle of helium is able to present two kinds of excitations. One of these motions is normal with the same properties as that of an ordinary viscous liquid, the other is superfluid ([3]).

The model that we propose is based on the interpretation of superfluidity as a second order phase transition, similar to the transition occurring in superconductivity. Indeed, as pointed out by several authors, there are many experimental phenomena (e.g., absence of dissipation, existence of a critical velocity, creation of vortices), showing the analogies between the behavior of superfluids and superconductors ([4]). In order to describe the phenomenon of superfluidity, we suggest a Ginzburg-Landau model, which initially was introduced in the framework of superconductivity, but has been recently

*Email: berti@dm.unibo.it
†Email: fabrizio@dm.unibo.it.

adopted to study phase transitions induced by temperature variations. The resulting differential system allows to explain the main aspects typical of superfluidity. Following the idea of the Cooper pairs in the context of superconductivity, we assume that each particle behaves as a couple of particles endowed with a normal and a superfluid excitation and assume that the normal component is related to the superfluid component by means of a constitutive equation similar to London's equation in superconductivity. Accordingly the evolution equation for the superfluid component is separated from the equation of the normal component.

Thermal effects are considered by introducing the evolution equation for the temperature deduced by the first principle of thermodynamics. In particular, we assume that the heat equation involves the velocity of the superfluid component. This model agrees with the second law of thermodynamics, since the Clausius-Duhem inequality is satisfied.

We analyze the well-posedness of the differential system constituted by the Ginzburg-Landau equation, by the evolution equation of the superfluid component and by the heat equation. For this problem we have proved existence and uniqueness of the solutions.

2 A Ginzburg-Landau model for superfluidity

Let us consider a superfluid occupying a bounded domain $\Omega \subset \mathbb{R}^3$. The passage from the normal to the superfluid state induces a change in the internal structure and it is characterized as a second order phase transition, since no latent heat is released. According to Landau's theory of phase transitions, the phase is represented by the scalar parameter $f \in [-1, 1]$, related to the density n_s of the superfluid particles by the formula

$$n_s = f^2.$$

Hence the condition $f^2 = 0$ denotes the normal state, while $f^2 \in (0, 1]$ describes a superfluid state. Since superfluidity is a second order phase transition, under the transition temperature, the phase of some particles can be still equal to zero, while some particles are in the superfluid phase.

The evolution of f is ruled by the Ginzburg-Landau equation typical of second order phase transitions,interpreted as a balance law of the internal order structure (see [2]), namely

$$f_t = \frac{1}{\kappa^2}\Delta f - f(f^2 - 1 + u + \lambda p + \mathbf{v}_s^2), \tag{1}$$

where $u = \theta/\theta_c$ is the normalized temperature, p is the pressure, $\mathbf{v}_s$ is the superfluid component of the velocity and κ, λ are positive constants.

The superfluid component $\mathbf{v}_s$ is assumed to satisfy the equations

$$\frac{\partial \mathbf{v}_s}{\partial t} = -\frac{\mu}{\kappa^2}\nabla\phi - \mu\nabla \times \nabla \times \mathbf{v}_s - \mu f^2 \mathbf{v}_s + \nabla u + \mathbf{h} \tag{2}$$

$$\nabla \cdot (f^2 \mathbf{v}_s) = -\kappa^2 f^2 \phi, \tag{3}$$

where ϕ is a suitable scalar function and the vector $\mathbf{h}$ is related to the body force. Equations (2) and (3) are similar to the evolution equations governing the motion of superconducting electrons, emphasizing the evident analogies between superfluidity and superconductivity.

We suppose that the normal component $\mathbf{v}_n$ is related to the superfluid component $\mathbf{v}_s$ through the constitutive equation

$$\mathbf{v}_n = \nabla \times \mathbf{v}_s. \tag{4}$$

Therefore the evolution equation for $\mathbf{v}_n$ is obtained by taking the curl of (2), namely

$$\frac{\partial \mathbf{v}_n}{\partial t} = -\mu \nabla \times \nabla \times \mathbf{v}_n - \mu \nabla \times (f^2 \mathbf{v}_s) + \nabla \times \mathbf{h}.$$

Equations (1)-(4) allow to explain some aspects related to the phenomenon of superfluidity.

First, we examine the phase diagram of ^{4}He, represented in figure 1. We notice that

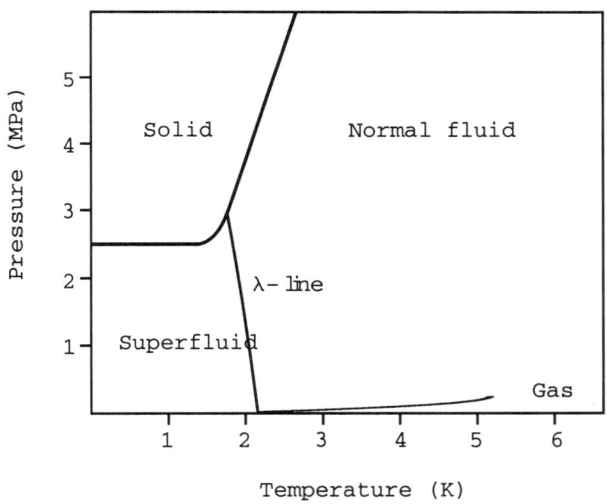

Figure 1: Phase diagram of ^{4}He

the curve separating the two phases, called λ–line, exhibits a high slope such that it can be approximated by a vertical line. By considering equation (1), we observe that if

$$R = u + \lambda p + \mathbf{v}_s^2 - 1 > 0,$$

then the fluid is in the normal state, since $f = 0$ is the solution of (1), with the boundary and initial conditions

$$\nabla f \cdot \mathbf{n}|_{\partial\Omega} = 0, \qquad f(x, 0) = 0.$$

Conversely, when $R < 0$ the pattern is able to describe the superfluid phase. Therefore, the phase transition occurs when

$$u + \lambda p + \mathbf{v}_s^2 = 1. \tag{5}$$

If we consider the equilibrium states, i.e. $\mathbf{v}_s = 0$, then (5) reduces to

$$u + \lambda p = 1,$$

which is a good approximation of the λ–line.

In addition, if $\mathbf{v}_s^2$ overcomes a threshold value, then $R > 0$, so that superfluidity is destroyed. This proves existence of a critical velocity, experimentally observed, above which superfluid properties disappear.

Henceforth, we choose $\lambda = 0$, which corresponds to the approximation of the λ–line with a vertical line. Then (1) reduces to

$$f_t = \frac{1}{\kappa^2}\Delta f - f(f^2 - 1 + u + \mathbf{v}_s^2).$$

Now we analyze the physical meaning of (2)-(4). The term ∇u accounts for the thermo-mechanical effect, according to which the particles of the superfluid flow in the same direction as the heat flux.

Since in (2) ∇u has the same sign of the acceleration of the superfluid component, the gradient of the temperature causes an increase of the superfluid component $\mathbf{v}_s$.

Moreover, suppose that a capillary is connected with a vessel containing the superfluid, and that owing to the narrowness of the tube the motion across the capillary satisfies $\nabla \times \mathbf{v}_s = 0$. Then, in view of (4),

$$\mathbf{v}_n = \nabla \times \mathbf{v}_s = 0,$$

so that the normal component is absent inside the tube. This means that only the superfluid component is allowed to flow across capillaries while the normal component undergoes a viscous resistance which forbids the flow to cross the tube.

2.1 Thermodynamical consistence of the model

The evolution equation for the temperature is deduced from energy balance law

$$E_t = \mathcal{P}^f + \mathcal{P}^{\mathbf{v}_s} + h\,, \tag{6}$$

where E is the total energy, h is the rate at which heat is absorbed and $\mathcal{P}^f, \mathcal{P}^{\mathbf{v}_s}$, are the internal powers related to the variables $f, \mathbf{v}_s$, defined as

$$\begin{aligned}
\mathcal{P}^f &= f_t^2 + \frac{1}{2\kappa^2}\left[(\nabla f)^2\right]_t + \frac{1}{4}\left[(1-f^2)^2\right]_t + f f_t(u + \mathbf{v}_s^2) \\
\mathcal{P}^{\mathbf{v}_s} &= \frac{1}{2}\left(\mathbf{v}_n^2\right)_t + \mu^{-1}\left(\frac{\partial \mathbf{v}_s}{\partial t} + \frac{\mu}{\kappa^2}\nabla\phi - \nabla u\right)^2 + f^2\mathbf{v}_s\cdot\frac{\partial \mathbf{v}_s}{\partial t} + u\nabla\cdot(f^2\mathbf{v}_s) \\
&\quad + \mu f^2\phi^2
\end{aligned}$$

We define the internal energy as

$$e = C(u) + \frac{1}{4}(1-f^2)^2 + \frac{1}{2\kappa^2}(\nabla f)^2 ,$$

where C is a function of the temperature such that $c(u) = C'(u)$ is the specific heat.

Moreover, we consider each particle of the superfluid as a pair endowed with two excitements represented by the components $\mathbf{v}_n$ and $|f|\mathbf{v}_s$. Accordingly, the kinetic energy of the particle is defined as

$$T = \frac{1}{2}(\mathbf{v}_n^2 + f^2\mathbf{v}_s^2).$$

The previous representation of the kinetic energy is in accordance with the representation of the velocity

$$\mathbf{v} = \mathbf{v}_n + |f|\mathbf{v}_s,$$

when the components $\mathbf{v}_n$ and $\mathbf{v}_s$ are orthogonal.

Hence, the total energy is given by

$$E = T + e = \frac{1}{2}\mathbf{v}_n^2 + \frac{1}{2}f^2\mathbf{v}_s^2 + C(u) + \frac{1}{4}(1-f^2)^2 + \frac{1}{2\kappa^2}(\nabla f)^2 ,$$

Substitution into (6) leads to

$$h = c(u)u_t - f_t^2 - \mu^{-1}\left(\frac{\partial \mathbf{v}_s}{\partial t} + \frac{\mu}{\kappa^2}\nabla\phi - \nabla u\right)^2 - uff_t - u\nabla\cdot(f^2\mathbf{v}_s) - \mu f^2\phi^2 . \quad (7)$$

The heat balance is expressed by the equation

$$h = -\nabla\cdot\mathbf{q} + r ,$$

from which we deduce the entropy equation

$$\frac{h}{u} - \frac{1}{u^2}\mathbf{q}\cdot\nabla u = -\nabla\cdot\left(\frac{\mathbf{q}}{u}\right) + \frac{r}{u} . \quad (8)$$

Here the heat flux $\mathbf{q}$ is supposed to obey the constitutive Fourier law

$$\mathbf{q} = -k(u)\nabla u , \quad (9)$$

where the conductivity k depends on the absolute temperature.

By substituting (7), (9) into (8) and neglecting the quadratic terms we obtain

$$\frac{c(u)}{u}u_t - ff_t = \nabla\cdot\left(\frac{k(u)}{u}\nabla u\right) + \nabla\cdot(f^2\mathbf{v}_s) + \frac{r}{u} . \quad (10)$$

Since the thermal conductivity and the specific heat are very small when the absolute temperature is close to zero, in this framework we will consider the linear approximations

$$k(u) = k_0 u, \qquad c(u) = c_0 u, \qquad k_0, c_0 > 0.$$

Moreover we assume that $\hat{r} = r/u$ is a known function of x, t. Accordingly (10) reduces to

$$c_0 u_t - f f_t = k_0 \Delta u + \nabla \cdot (f^2 \mathbf{v}_s) + \hat{r} \, . \tag{11}$$

We show that equations (2), (3) and (11) allow to prove the existence of second sound effect, revealing that superfluids are able to propagate thermal waves. For this problem, let us suppose that f is constant and that $\mathbf{h}, \hat{r}$ vanish. Therefore the heat equation (11) reads

$$c_0 u_t = k_0 \Delta u + f^2 \nabla \cdot \mathbf{v}_s. \tag{12}$$

Besides the heat flux due to the Fourier conduction law, we observe that (12) contains a flux proportional to the superfluid component $\mathbf{v}_s$. In addition, the velocity $\mathbf{v}_s$ satisfies the equation

$$\frac{\partial \mathbf{v}_s}{\partial t} = -\frac{\mu}{\kappa^2} \nabla \phi - \mu \nabla \times \nabla \times \mathbf{v}_s - \mu f^2 \mathbf{v}_s + \nabla u, \tag{13}$$

which has a similar structure of the heat propagation law in the Cattaneo-Maxwell regime. Since the heat propagation in a superfluid is not primarily due to the thermal conductivity of the material, we suppose $k_0 \ll 1$. Then, the effect of the superfluid component is prevailing, so that the second sound is expected. Indeed, (3) and (13) lead to

$$(\nabla \cdot \mathbf{v}_s)_t = -\frac{\mu}{\kappa^2} \Delta \phi + \mu \kappa^2 f^2 \phi + \Delta u$$

Therefore, by neglecting thermal conductivity, from (12) we deduce

$$c_0 u_{tt} - f^2 \Delta u = f^2 (\kappa^2 \mu f^2 \phi - \frac{\mu}{\kappa^2} \Delta \phi),$$

which means that the temperature satisfies an equation able to provide a new kind of wave.

We conclude this section, showing that the model proposed is consistent with the second law of thermodynamics. To this purpose, let us write the Clausius-Duhem inequality in the non-local formulation

$$\frac{\partial \eta}{\partial t} \geq \frac{h}{u} - \frac{1}{u^2} \mathbf{q} \cdot \nabla u + \nabla \cdot \mathbf{N}. \tag{14}$$

where η is the entropy and $\mathbf{N}$ is a suitable flux vector such that $\mathbf{N} \cdot \mathbf{n}|_{\partial \Omega} = 0$. In view of the definitions (7), (9) of $h, \mathbf{q}$ and with the choices

$$\eta(u, f) = c_0 u - \frac{1}{2} f^2, \qquad \mathbf{N} = f^2 \mathbf{v}_s,$$

inequality (14) implies

$$0 \geq -f_t^2 - \mu^{-1} (\mathbf{v}_{st} + \nabla \phi - \nabla u)^2 - \mu f^2 \phi^2 \, ,$$

which proves the thermodynamical consistence of the model.

2.2 Rotation of a superfluid

Liquid helium has a peculiar behavior when it is set into rotation. Let us consider a cylindrical bucket of radius R, containing Helium II rotating around its axis with angular velocity ω. When ω exceeds a critical value ω_{c_1}, parallel straight vortex lines appear and their number grows with the increasing of the angular velocity. However, if ω overcomes a second threshold value ω_{c_2}, the cores of the vortices overlaps and superfluidity is destroyed.

The phenomenon is identical to the creation of vortices observed by Abrikosov in a type II superconductor: if the intensity of an impressed magnetic field is between two critical values H_{c1} and H_{c2}, vortex lines occur. By letting $\mathbf{h} = 0$, the stationary problem in isothermal conditions is ruled by the system

$$\frac{1}{\kappa^2}\Delta f - f(f^2 - 1 + u + \mathbf{v}_s^2) = 0 \tag{15}$$

$$-\frac{1}{\kappa^2}\nabla\phi - \nabla\times\nabla\times\mathbf{v}_s - f^2\mathbf{v}_s = 0 \tag{16}$$

$$\nabla\cdot(f^2\mathbf{v}_s) + \kappa^2 f^2\phi = 0 \tag{17}$$

We associate the boundary conditions

$$\nabla f\cdot\mathbf{n}|_{\partial\Omega} = 0\,, \qquad (\nabla\times\mathbf{v}_s)\times\mathbf{n}|_{\partial\Omega} = \omega\,, \qquad \mathbf{v}_s\cdot\mathbf{n}|_{\partial\Omega} = 0. \tag{18}$$

Since (15)-(18) coincide with the Ginzburg-Landau equations studied by Abrikosov for superconductors of type II, with the same analysis, we may prove the existence of vortex lines in a superfluid.

We restrict our attention to the case $f^2 = \lambda^2 = const$. Due to the symmetry of the domain, we introduce a system of cylindrical coordinates (r, θ, z) with versors of the axes $\mathbf{e}_r, \mathbf{e}_\theta, \mathbf{e}_z$ respectively and look for solutions of (16)-(17) with

$$\mathbf{v}_s = v_s(r)\mathbf{e}_z.$$

Then v_s satisfies the differential equation

$$v_s'' + \frac{1}{r}v_s' - \lambda^2 v_s = 0, \tag{19}$$

with boundary condition

$$v_s'(R) = \omega.$$

Equation (19) is a Bessel equation, whose bounded solution is

$$v_s(r) = \frac{\omega}{\lambda I_1(\lambda R)} I_0(\lambda r),$$

where I_0, I_1 are the modified Bessel functions of the first kind.

Moreover the normal component is

$$\mathbf{v}_n = \nabla\times\mathbf{v}_s = -v_s'(r)\mathbf{e}_\theta = v_n(r)\mathbf{e}_\theta,$$

with

$$v_n(r) = -\frac{\omega}{I_1(\lambda R)} I_1(\lambda r) \to 0, \qquad \text{as } r\to 0.$$

3 Well-posedness of the problem

The model proposed to describe superfluidity considers as state variables the order parameter f, the superfluid component $\mathbf{v}_s$ and the absolute temperature θ, satisfying the differential system

$$\begin{aligned} f_t &= \frac{1}{\kappa^2}\Delta f - f(f^2 - 1 + u + \mathbf{v}_s^2) \\ (\mathbf{v}_s)_t &= -\nabla\phi - \mu\nabla\times\nabla\times\mathbf{v}_s - f^2\mathbf{v}_s + \nabla u + \mathbf{h} \\ c_0 u_t - f f_t &= \nabla\cdot(f^2\mathbf{v}_s + k_0\nabla u) + r \\ \nabla\cdot(f^2\mathbf{v}_s) &= -\kappa^2 f^2\phi \end{aligned}$$

We associate the boundary and initial conditions

$$\begin{aligned} &\nabla f\cdot\mathbf{n}|_{\partial\Omega} = 0\,, \quad (\nabla\times\mathbf{v}_s)\times\mathbf{n}|_{\partial\Omega} = \boldsymbol{\omega}\,, \quad \mathbf{v}_s\cdot\mathbf{n}|_{\partial\Omega} = 0, \quad u|_{\partial\Omega} = u_b \\ &f(x,0) = f_0(x), \quad \mathbf{v}_s(x,0) = \mathbf{v}_{s0}(x), \quad u(x,0) = u_0(x)\,. \end{aligned}$$

Existence and uniqueness of solutions are proved by means of a suitable decomposition of the unknown $\mathbf{v}_s, \phi$, i.e.

$$\mathbf{v}_s = -\mathbf{A} + \frac{1}{\kappa}\nabla\varphi, \qquad \phi_s = \nabla\cdot\mathbf{A} - \frac{1}{\kappa}\varphi_t, \qquad \mathbf{A}\cdot\mathbf{n}|_{\partial\Omega} = 0\,.$$

By letting $\psi = fe^{i\varphi}$, the problem can be written by means of the variables $\psi, \mathbf{A}, u$ in the form

$$\begin{aligned} \psi_t &= \frac{1}{\kappa^2}\Delta\psi - \frac{2i}{\kappa}\mathbf{A}\cdot\nabla\psi - \psi|\mathbf{A}|^2 + i(\kappa - \frac{1}{\kappa})(\nabla\cdot\mathbf{A})\psi \\ &\quad -\psi(|\psi|^2 - 1 + u) && (20) \\ \mathbf{A}_t &= \nabla(\nabla\cdot\mathbf{A}) - \mu\nabla\times\nabla\times\mathbf{A} - |\psi|^2\mathbf{A} + \frac{i}{2\kappa}(\psi\nabla\bar\psi - \bar\psi\nabla\psi) \\ &\quad -\nabla u - \mathbf{h} && (21) \\ c_0 u_t &= \frac{1}{2}(\psi_t\bar\psi + \psi\bar\psi_t) + k_0\Delta u \\ &\quad +\nabla\cdot\left[-|\psi|^2\mathbf{A} + \frac{i}{2\kappa}(\psi\nabla\bar\psi - \bar\psi\nabla\psi)\right] + \hat r && (22) \end{aligned}$$

with boundary and initial conditions

$$\begin{aligned} &\nabla\psi\cdot\mathbf{n}|_{\partial\Omega} = 0, \quad (\nabla\times\mathbf{A})\times\mathbf{n}|_{\partial\Omega} = -\boldsymbol{\omega}, \quad \mathbf{A}\cdot\mathbf{n}|_{\partial\Omega} = 0, \quad u|_{\partial\Omega} = u_b && (23) \\ &\psi(x,0) = \psi_0(x) \qquad \mathbf{A}(x,0) = \mathbf{A}_0(x) \qquad u(x,0) = u_0(x). && (24) \end{aligned}$$

For sake of simplicity, $\mathbf{h}, \hat r, \boldsymbol{\omega}$ and u_b are assumed to be time independent.

Theorem 1. *Let $\psi_0, \mathbf{A}_0 \in H^1(\Omega), u_0 \in L^2(\Omega)$, $\mathbf{h}, \hat r \in L^2(\Omega)$. Then for every $T > 0$, there exists a unique solution $(\psi, \mathbf{A}, u)$ to (20)-(24) such that*

$$\begin{aligned} \psi,\ \mathbf{A} &\in L^2(0,T,H^2(\Omega)) \cap H^1(0,T,L^2(\Omega)) \\ u &\in L^2(0,T,H^1(\Omega)) \cap H^1(0,T,H^1(\Omega)')\,. \end{aligned}$$

Moreover ψ, $\mathbf{A} \in C(0,T,H^1(\Omega))$, $u \in C(0,T,L^2(\Omega))$.

Existence of a solution can be established, first by proving that system (20)-(24) admits at least a solution defined in a small time interval $[0;\tau)$. Next, by means of uniform estimates, such a solution can be extended in the whole time interval $[0;T)$ for every $T > 0$. Uniqueness is proved with standard techniques. The details of the proof of Theorem 1 can be found in [1].

References

[1] V. Berti, M. Fabrizio, Existence and uniqueness for a mathematical model in superfluidity, *Math. Meth. Appl. Sci*, DOI: 10.1002/mma.981.

[2] Fabrizio M. Ginzburg-Landau equations and first and second order phase transitions. *Internat. J. Engrg. Sci.* **44**, (2006), 529-539.

[3] Landau LD, Lifshitz EM. *Fluid Mechanics*, Pergamon Press, Oxford, 1987.

[4] Mendelssohn K. *Liquid Helium*, in S. Flugge (ed.), Handbuch Physik, Springer, Berlin, 1956, 370-461.

Spherical Tolman-Bondi Models in Cosmology

I. Bochicchio[*], *E. Laserra*[†]

Department of Mathematics and Informatics,
University of Salerno,
via Ponte Don Melillo, 84084 Fisciano (SA), Italy.

Abstract

Spherical symmetry is considered and exact solutions of Tolman–Bondi equations are studied taking advantage from Ricci principal curvature depending on the radial coordinate. Moreover an expansion of the exact solutions in fractional Puiseux series in considered to compare Euclidean and not Euclidean cases.

1 Introduction

In this paper we analyze the Tolman–Bondi equations to describe various properties of spherically symmetrical composed of dust particles which exert negligible pressure and which are distributed non uniformly but nevertheless with spherical symmetry around some particular origin. More precisely we give the following assumption [2]:

- the system is and remains spherically symmetric, i.e. the mass density and particle velocity are functions of a radial coordinate r and a time coordinate t only;
- each particle moves under the influence of gravity only. This implies that there are no electromagnetic force acting on the particles and that there are no pressures;
- the orbit of particle do not intersect [1] and the mass density is everywhere finite.

In addition to these assumptions we will also put the cosmological constant $\lambda = 0$.

It follows that, called t the proper time of of each particle and $(r,\ \theta,\ \varphi)$ the co–moving spherical coordinates, our metric is [2]

$$\mathrm{d}\,s^2 = A^2(t,r)\mathrm{d}\,r^2 + B^2(t,r)(\mathrm{d}\,\theta^2 + \sin^2\theta d\varphi^2) - \mathrm{c}^2\,\mathrm{d}\,t^2\ , \tag{1}$$

where $B(t,r)$ can be interpreted as the intrinsic radius, at time t, of the geodesic sphere $\mathcal{S}(r)$ with its center at the centre of symmetry O and co-moving radius r [6].

*Email: ibochicchio@unisa.it

†Email: elaserra@unisa.it

[1]The exclusion of intersecting orbits does not lack physical meaning since, if they intersect, pressures would certainly arise.

We consider now the initial space-like hypersurface V_3 (with equation $t = 0$) and call r–shells the set of particles with co–moving radius r (i.e. the dust initially distributed on the surface of the geodesic sphere $\mathcal{S}(r)$). In accordance with [4, 5] we assign each particle of an r–shell the initial intrinsic radius $r = B(r, 0)$ as radial co–moving coordinate. Hence the metric of the initial spatial manifold V_3 takes the form

$$\mathrm{d}\,\sigma^2 = a^2(r)\mathrm{d}\,r^2 + r^2(\mathrm{d}\,\theta^2 + \sin^2\theta\mathrm{d}\,\varphi^2)\ , \tag{2}$$

where $a(r) = A(r, 0)$. Then the Bondi evolution equations [2] take the form [4, 5]: [2]

$$\begin{cases} \dot{B}^2(t,r) &=& \mathrm{c}^2\left(\dfrac{1}{a^2(r)} - 1\right) + \dfrac{2\,G_N\,M(r)}{B(t,r)} \\[2ex] A(t,r) &=& a(r)\,B'(t,r) \\[2ex] \mu(t,r) &=& \dfrac{\mu_0(r)r^2}{B'(t,r)B^2(t,r)} \end{cases} \tag{3}$$

where $M(r)$ is the so–called *euclidean mass* [4, 5] [3]

$$M(r) = 4\pi\int_0^r \mu_0(s)\,s^2\,\mathrm{d}\,s\ , \tag{4}$$

and $\mu(t, r)$ is the mass density ($\mu_0 = \mu(r, 0)$).

By means of the first principal curvature and an expansion in fractional Puiseux series, we are going to analyze the solutions of the field equations.
Moreover we take advantage of the mechanical analogy between the relativistic evolution of a spherical dust universe and the classical unidimensional motion of bodies freely falling toward or from a newtonian center of attraction. To this aim we introduce a suitable function $\varepsilon(r)$ that plays the role of the mechanical energy in the motion of a freely falling body in Classical Mechanics, so that we are able to qualitatively study the evolution of the spherical universe through a method analogous to the classical energy method.

2 The principal curvatures in spherical symmetry and solutions of evolution equations

To determinate the influence of the initial spatial geometry on the subsequent evolution of a spherical dust universe, in this section we shall consider the curvature properties of a spherically symmetric V_3[4] by the method introduced by Ricci (valid for a generic V_3 only, see [6, Chapter XII §11]), taking into account that the metric $\mathrm{d}\,\sigma^2 = \gamma_{ij}\,\mathrm{d}\,x^i\,\mathrm{d}\,x^j$

---

[2] Hereafter a dot will denote differentiation with respect to t and a prime differentiation with respect to r.

[3] If the initial spatial manifold V_3 is euclidean, then $M(r)$ is the mass of dust contained within $\mathcal{S}(r)$.

[4] Most of the following considerations are valid for a generic spherically symmetric strictly–Riemannian manifold V_3, not necessarily limited to the case of a spatial initial manifold of a dust universe.

of a spherically symmetric manifold V_3 can be given under the form (2).
More precisely, the curvature properties of a 3-dimensional manifold V_3 are completely determined by a double symmetric tensor α_{ij}, the Ricci curvature tensor of V_3 (see [6, Chap.VII §11 p.198–201])

$$\alpha^{ij} = \frac{1}{4}\,\epsilon^{ipq}\epsilon^{jrs}R_{pqrs}\ , \tag{5}$$

where ϵ^{ijk} is the standard anti–symmetric unit tensor. In fact, the three principal curvatures $\omega_i\,,\ i\ =\ 1,2,3,$ are defined as the roots of the characteristic equation $\det||\alpha_{ij} - \omega\,\gamma_{ij}|| = 0$ [6, Chap.VII §12 (43) p.205], so they are the principal invariants determined by the Ricci tensor α_{ij}. Moreover they have the following expression [6, Chapter XII §11 p.412–414]:

$$\omega_1 = \frac{1}{r^2}\left(1 - \frac{1}{a(r)^2}\right)\,, \quad \omega_2 = \omega_3 = \frac{1}{2\,r}\frac{\mathrm{d}}{\mathrm{d}\,r}\left(1 - \frac{1}{a(r)^2}\right) \equiv \omega\ . \tag{6}$$

Note that the geometry of V_3 is completely determined by the principal curvature ω_1 only. In particular the curvature properties of V_3 are also completely determined by ω_1 alone.

Now we will focus our attention on a given single r-shell (that is we will consider r as a temporarily fixed parameter); so we can regard the intrinsic radius B as a function of time only, $B = B(t;r)$, and $\varepsilon\,,\omega_1\,,M,\ldots$ as constants.
As such, introducing the first principal curvature ω_1 into the evolution equation $(3)_1$, it becomes a non–linear ordinary differential equation [5]

$$\dot{B}^2 = -\omega_1\,r^2\,\mathrm{c}^2 + \frac{2\,G_N M(r)}{B}\ . \tag{7}$$

By introducing the function [6]

$$R_S(r) = \frac{2\,G_N\,M(r)}{\mathrm{c}^2}\ ,$$

we can give the equation (7) the simple form

$$\dot{B}^2 = \mathrm{c}^2\left(-\omega_1\,r^2 + \frac{R_S(r)}{B}\right)\ . \tag{8}$$

It is evident from Eq. (8) that the behavior of the solutions is determined by the sign of ω_1, so in the following we will put $\varpi = |\omega_1|$ and we will study this equation in the three different cases $\omega_1 = 0$, $\omega_1 = \varpi > 0$ and $\omega_1 = -\varpi < 0$, to get the corresponding exact solutions.

The case $\omega_1\ =\ 0$ corresponds to the spatially Euclidean case $a^2(r)\ =\ 1$, so the intrinsic radius coincides with the ordinary radius, and it is the only case where it is possible to solve equation (8) explicitly for B: [7]

$$t - T_e = \pm\frac{2\,B^{\frac{3}{2}}}{3\,\mathrm{c}\,\sqrt{R_s(r)}} \Rightarrow B(t;r) = r\left(1 - \frac{t}{T_e}\right)^{2/3}\ . \tag{9}$$

[5] Hereafter a dot will denote total differentiation with respect to t, $\dot{B} = \frac{\mathrm{d}\,B}{\mathrm{d}\,t}$

[6] $R_S(r)$ is formally the Schwarschild radius associated with the Euclidean mass $M(r)$ of the generic r–sphere.

[7] Actually T_e is an arbitrary function $T_e(r)$ such that for $t = T_e(r) \Rightarrow B = 0$. So $T_e(r)$ is the time at which the dust distributed on the r-shell is collapsed into the symmetry center.

If $\omega_1 = \varpi > 0$, equation (8) becomes

$$\left(\frac{\mathrm{d}\,B}{\mathrm{d}\,t}\right)^2 = \mathrm{c}^2\left(-\varpi\, r^2 + \frac{R_S}{B}\right)\,. \tag{10}$$

By introducing the new function

$$B_M(r) = \frac{R_S(r)}{\varpi(r)\, r^2} \tag{11}$$

we can write equation (10) as

$$\left(\frac{\mathrm{d}\,B}{\mathrm{d}\,t}\right)^2 = \mathrm{c}^2\,\varpi\, r^2\left(\frac{B_M}{B} - 1\right) \tag{12}$$

so $B_M(r) = B_M$ is the maximum value that the intrinsic radius of the given r–shell can reach. [8]

In this case we obtain the temporal law of evolution in the following form:

$$t - T_p = \frac{\pm 1}{\mathrm{c}\, r\sqrt{\varpi}}\left(B_M \arctan\sqrt{\frac{B}{B_M - B}} - \sqrt{B(B_M - B)}\right)\,, \tag{13}$$

where T_p represents the times $T(r)$ where the corresponding r–shells are collapsed into the symmetry centre.

It is not possible to expand the second member of (13) in a power series with respect to B, but we can expand it in fractional Puiseux series [9]

$$\begin{aligned}
&\frac{1}{\mathrm{c}\, r\sqrt{\varpi}}\left(B_M \arctan\sqrt{\frac{B}{B_M - B}} - \sqrt{B(B_M - B)}\right) = \\
=&\frac{1}{\mathrm{c}\, r\sqrt{\varpi}}\left(\frac{2\,B^{\frac{3}{2}}}{3\sqrt{B_M}} + \frac{B^{\frac{5}{2}}}{5\,B_M^{\frac{3}{2}}} + \frac{3\,B^{\frac{7}{2}}}{28\,B_M^{\frac{5}{2}}} + \ldots\right) = \\
=&\frac{2}{3\,\mathrm{c}\sqrt{R_s}}\,B^{\frac{3}{2}} + \cdots\cdots
\end{aligned} \tag{14}$$

to show that *the spatially Euclidean solution* (9) *corresponds to the first term of the expansion of the spatially non–Euclidean solution* (13) *in fractional Puiseux series.*

Finally, if $\omega_1 = -\varpi < 0$, we can write equation (8) as

$$\left(\frac{\mathrm{d}\,B}{\mathrm{d}\,t}\right)^2 = \mathrm{c}^2\,\varpi\, r^2\left(1 + \frac{B_M}{B}\right)\,, \tag{15}$$

where B_M is still equal to $\frac{R_S}{\varpi}$ but, unlike (11), it does not represent a maximum of the function $B(t;r)$. In this case we obtain the following temporal law of evolution:

$$t - T_n = \frac{\pm 1}{\mathrm{c}\, r\sqrt{\varpi}}\left[\sqrt{B\,(B + B_M)} - B_M \operatorname{arsinh}\left(\frac{\sqrt{B}}{\sqrt{B_M}}\right)\right]\,, \tag{16}$$

---

[8] In fact we can write (12) as $\left(\frac{\mathrm{d}\,B}{\mathrm{d}\,t}\right)^2 = R_S\left(\frac{1}{B} - \frac{1}{B_M}\right) \Rightarrow B(t;r) \leq B_M(r)$.

[9] We can develop it in Mac Laurin series with respect to $\sqrt{B}$ thus obtaining a fractional power series. As it is known the fractional power series are particular cases of Puiseux series (see e.g. [1]).

where T_n represents the times $T(r)$ where the corresponding r–shells are collapsed into the symmetry centre.
It is not possible to expand the second member of (16) in a power series with respect to B, but we can again expand it in fractional Puiseux series:

$$\begin{aligned}
&\frac{1}{c\,r\,\sqrt{\varpi}}\left[\sqrt{B\,(B+B_M)}-B_M\,\operatorname{arsinh}\left(\frac{\sqrt{B}}{\sqrt{B_M}}\right)\right]=\\
=&\frac{1}{c\,r\,\sqrt{\varpi}}\left[\frac{2\,B^{\frac{3}{2}}}{3\,\sqrt{B_M}}-\frac{B^{\frac{5}{2}}}{5\,B_M^{\frac{3}{2}}}+\frac{3\,B^{\frac{7}{2}}}{28\,B_M^{\frac{5}{2}}}+\cdots\right] \qquad (17)\\
=&\frac{2}{3\,c\,\sqrt{R_s}}\,B^{\frac{3}{2}}+\cdots\cdots
\end{aligned}$$

to show that *the spatially Euclidean solution* (9) *again corresponds to the first term of the expansion of the spatially not–Euclidean solution* (16) *in fractional Puiseux series.*

Hence, in general, the evolution of a spatially spherical dust universe depends on the initial spatial curvature, but if we consider an expansion of the exact solutions of evolution equations in fractional Puiseux series it is possible shown [3] that, in proximity of the collapse/generation times, a generic spatially spherical dust universe evolves analogously to the spatially euclidean case. Hence, it seems that the first principal curvature is uninfluence in suitable sufficiently *small* neighbor of the symmetry center.

3 Qualitative evolution of a relativistic r-shell

To qualitatively study the evolution of the spherical dust universes, we are going to introduce the function [10]

$$\varepsilon(r)=\frac{1}{a^2(r)}-1 \qquad (18)$$

into equation $(3)_1$

$$\dot{B}^2(r,t)=\varepsilon(r)\,c^2+\frac{2\,G_N\,M(r)}{B(r,t)}\,. \qquad (19)$$

Now we will focus our attention on a given single r-shell, by considering fixed the variable r, which becomes a parameter. So equation (19) becomes

$$\dot{x}^2=\varepsilon\,c^2+\frac{2\,G_N\,M}{x} \qquad (20)$$

and takes the form of the energy integral in classical mechanics. In fact let's identify

ε with twice the ratio between the classical energy E and the relativistic intrinsic energy of a falling body of unit mass, $\varepsilon=\frac{2\,E}{c^2}$; [11]

$M(r)$ (the euclidean mass) with the mass of an attracting material point *fixed* in O;

---

[10] The function $|\varepsilon(r)|$ represents the percentage deviation of $a^2(r)$ from the Euclidean value $a^2=1$.

[11] For simplicity we will consider freely falling bodies of unit mass, but, if we prefer to consider falling bodies of mass m, we can obviously put $\varepsilon=\frac{2\,E}{m\,c^2}$.

$x(t) = B(t; r)$ (the intrinsic radius of the given r–shell) with the distance from O of the falling body;

$\dot{x}(t) = \dot{B}(t; r)$ with the velocity of the falling body; [12]

then equation (20) becomes formally analogous to the energy integral for a falling body of unit mass in a newtonian force field

$$\dot{x}^2 = 2\,E + \frac{2\,G_N\,M}{x}\,, \tag{21}$$

where E is the mechanical energy and $-\frac{G_N\,M}{x}$ is the potential energy $V(x)$. Hence in the following, on the analogy of the classical falling bodies, we could call *energy* the function $E(r) = \frac{c^2\,\varepsilon(r)}{2}$ and *reduced energy* the function $\varepsilon(r)$.

As it is known (see e.g. [8, p. 37–39]), we can easily obtain the qualitative behaviour of the solutions of equation (21) without integrate, through the classical energy method; [13] so we want to study evolution equation (20) by the *energy method*, on the analogy of the classical case.

For a falling body, the potential energy $V(x)$ is a branch of hyperbola with the positive x–semi–axis and the negative y–semi–axis as asymptotes. So if the mechanical energy $E > 0$ there is no turning point, if $E = 0$ there is a turning point at infinity and finally if $E < 0$ there is a finite turning point (see e.g. [8, p. 38]).

Remark 1. *As it is known* (see e.g. [7, p.210–211]), *apart from the sign of E, if the body is initially falling toward O, it will continue falling toward O, without inverting its motion and with an increasing velocity, as far as it reaches O after a finite time. Analogously if a r–shell is initially contracting, it will continue to contract toward the symmetry center O, apart from the sign of ε.*

Positive *energy*

When the mechanical energy $E > 0$, for a falling body there is no turning point. If the body is initially moving upward ($\dot{x}_0 > 0$), it will continue to move upward forever, with decreasing velocity, approaching the limit value $\dot{x}_l = \sqrt{2\,E}$.

Analogously when $\epsilon > 0$ if the intrinsic radius of the corresponding r–shell is initially increasing, it will continue increase forever, with decreasing rate, approaching the limit rate $\dot{B}_l = \sqrt{\varepsilon}\,c^2$.

Zero *energy*

When the mechanical energy $E = 0$, equation (20) becomes

$$\dot{x}^2 = \frac{2\,G_N M}{x} \tag{22}$$

[12] As usual the x–axis will be oriented from O to infinity, so the plus sign for the velocity $\dot{x}$ will refer to ascending motion, the minus sign to descending motion.

[13] For a given E the unidimensional motion is confined to those regions on the x-axis where $E \geq V(x)$. These regions are unlimited or are delimitated by the *stop* points x_S where the line of the energy $y = E$ intersects the diagram of potential energy $y = V(x)$, that is $E = V(x_S)$, and the velocity is null. In the case of a falling body, at the eventual stop points x_S the body stops and reverses its velocity (see e.g. [8, p. 33]); hence these points are called *turning points* x_T.

and the turning point for a falling body goes to infinity. If the body is initially moving upward ($\dot{x}_0 > 0$), it will move upward forever approaching the limiting velocity $v_l = 0$. At any height x the velocity will be the *escape velocity*

$$v_e = \sqrt{\frac{2\,G_N M}{x}}\ . \tag{23}$$

Analogously, when $\epsilon = 0$, for the intrinsic radius of the corresponding r–shell the *turning point* goes to infinity, and if it is initially increasing, it will continue expanding forever, approaching the limiting expansion rate $\dot{B}_l = 0$. The rate $\dot{B} = \dot{x}$ will be

$$\dot{B}_e = \sqrt{\frac{2\,G_N\,M}{B}}\ , \tag{24}$$

that, on the analogy of a falling body, we could consider *escape velocity* for a r–shell of intrinsic radius B.

Negative *energy*

As it is known (see e.g. [8, p.38]), when the mechanical energy $E < 0$, a falling body comes up to a turning point at a distance [14]

$$x_T = \frac{G_N\,M}{-E} \tag{26}$$

from O.

Analogously when $\varepsilon < 0$ the intrinsic radius of the corresponding r-shell comes up to a *turning point* [15]

$$B_T = -\frac{2\,G_N\,M}{\varepsilon\,c^2}\ . \tag{27}$$

If the r–shell is initially expanding, it will continue expanding until it reaches the maximum expansion point B_T then it will contract back toward the center of symmetry O until it collapses.

References

[1] Amerio L., *Analisi Matematica con elementi di Analisi Funzionale*, Vol. 3, Parte I, Utet.

[2] Bondi H., *Spherically Symmetrical Models in General Relativity*, Monthly Notices, **107** (1947), pp. 410–424.

[14] Equation (20) becomes

$$\dot{x} = \pm\sqrt{2\,G_N\,M\left(\frac{1}{x} - \frac{1}{x_T}\right)}\ , \tag{25}$$

if the body is initially moving upward, it will come to a stop at x_T, then it will fall back from x_T to O in a finite time.

[15] For an r-shell $x_T = B_T$ represents a point of maximum of the function $x(t) = B(t;r)$ that characterizes its evolution.

[3] Bochicchio I. and Laserra E., *Evolution of a Spherical Universe in a Short Range Collapse/Generation Interval*, Springer-Verlag Berlin Heidelberg, Y. Shi et al. (Eds.): ICCS 2007, Part II, LNCS 4488, (2007) pp. 997-1003.

[4] Laserra E., *Sul problema di Cauchy relativistico in un universo a simmetria spaziale sferica nello schema materia disgregata*, Rendiconti di Matematica, **2** (1982), pp. 799–816.

[5] Laserra E., *Dust universes with spatial spherical symmetry and euclidean initial hypersurfaces*, Meccanica, **20** (1985), pp. 267–271.

[6] Levi-Civita T., *The Absolute Differential Calculus*, Dover Publications,Inc., 1977.

[7] Levi-Civita T. and Amaldi U., *Lezioni di Meccanica Razionale*, Vol. 2, Part 1, Zanichelli, 1974.

[8] Symon K.R., *Mechanics*, Addison–Wesley Publishing Company, third edition (1971).

[9] Tolman R.C. *Effect of Inhomogeneity on Cosmological Models*, Proc. Nat. Acad. Sci., **20** (1934), pp 169–176.

Nonlinear Stability of a SIRS Epidemic Model with Convex Incidence Rate

B. Buonomo[*], *S. Rionero*[†]

Department of Mathematics and Applications,
University of Naples *Federico II*
via Cintia, 80126 Naples, Italy.

Abstract

We study an epidemic model for infections with non permanent acquired immunity (SIRS). The incidence rate is assumed to be convex respect to the infective class. By using a peculiar Lyapunov function, we obtain necessary and sufficient conditions for the local nonlinear stability of equilibria. Conditions ensuring the global stability of the endemic equilibrium are also obtained. Our procedure allows to enlarge the class of incidence rates ensuring the Lyapunov nonlinear stability of the endemic equilibrium for SIRS models.

Keywords: epidemic model, direct Lyapunov method, Lyapunov function, convex incidence rate.

1 Introduction

Recently, the following SIRS model, with a general incidence rate, $f(S,I)$, has been considered by A. Korobeinikov, [6], [7]:

$$\begin{cases} \dot{S} = \mu - f(S,I) - \mu S + \gamma R \\ \dot{I} = f(S,I) - (\sigma + \mu)I \\ \dot{R} = \sigma I - (\mu + \gamma)R, \end{cases} \tag{1}$$

where the upper dot denotes the time derivative, $d\cdot/dt$. The state variables are the fractions in which is divided a host population: the suceptibles, S, the infective, I, and the recovered, R. The parameters (all positive constants) have the following meaning: μ is the birth/death rate, σ is the recovery rate and γ is the rate at which recovered individuals lose immunity and return to the susceptible class.
The nonlinear incidence, f, is required to satisfy the following general assumptions:

$$\begin{cases} f : (S,I) \in Q \rightarrow f(S,I) \in \mathbf{R}_+, \\ f \in C^1(Q), \\ f(S,0) = f(0,I) = 0; \quad \text{for all } S > 0, I > 0. \end{cases} \tag{2}$$

[*]Email: buonomo@unina.it
[†]Email: rionero@unina.it

where $Q = [0,1] \times [0,1] \subset \mathbf{R}^2_+$. (for the general theory about epidemic models see e.g., [1, 3, 4, 5]).
By using the Lyapunov Direct Method, A. Korobeinikov found sufficient conditions on $f(S,I)$ ensuring that there exists an unique endemic equilibrium and it is globally asymptotically stable. Such conditions are satisfied by several popular incidence rates as, e.g., the ones reflecting the effect of saturation with respect to the infectives I. However, the conditions rule out some other incidence rates, as the ones which are convex (i.e. concave up) respect to the variable I. An example is the following one, which has been considered, e.g. in [14], [15]:

$$f(S,I) = kIS(1+\alpha I), \tag{3}$$

where k and α are positive constants. As stressed in [14], the use of this functional means the inclusion of an increased rate of infection due to two exposures over a short time period. The single contacts lead to infection at the rate kIS, whereas the new infective individuals arise from double exposures at a rate $\alpha k I^2 S$.
In this paper, we will perform a stability analysis for system (1), with incidence rate (3). We introduce a Lyapunov function, which is the ODEs 'adaptation' of a function first introduced by S.R. in the context of L^2-stability analysis for binary reaction-diffusion systems (see e.g. [10]-[13]). In our analysis, as in [6], we assume the equality between birth rate and death rate. This allows a better comparison with the result obtained in [6], although this restriction is unnecessary for our analysis to work.
We will find exact coincidence between the linear and non! linear stability of equilibria (at least locally) and a global stability result for the endemic equilibrium will be also obtained. Hence, we enlarge the class of incidence rates ensuring the nonlinear stability for model (1).

The paper is organized as follows. Some results concerning the SIRS model (1)-(3) are recalled in Section 2. The main result, i.e. the nonlinear stability/instability theorem, is contained in Section 3. In Section 4 sufficent conditions for the global stability are derived. Concluding remarks, in Section 5, close the paper.

2 Equilibria

Introduce the *basic reproductive rate*: $R_0 = k(\sigma+\mu)^{-1}$. Denoting by $N(t)$ the number of the total population, i.e. setting $N = S+I+R$, from (1) we get: $\dot{N} = \mu - \mu N$, so that for any initial $N(t_0) \geq 0$ it follows: $\lim_{t\to\infty} N(t) = 1$. Hence, the plane $S+I+R=1$, is an invariant manifold of system (1)-(3), which is attracting in the first octant, [15]. Here, we assume that the population is in equilibrium and investigate the dynamics of system (1)-(3) on such plane.

Equilibria $E(S^*, I^*, R^*)$ of system (1)-(3) satisfy the system:

$$\begin{cases} kS^*I^*(1+\alpha I^*) + \mu S^* = \mu + \gamma R^* \\ kS^*I^*(1+\alpha I^*) = (\sigma+\mu)I^* \\ \sigma I^* = (\mu+\gamma)R^* \\ S^* = 1 - I^* - R^*, \end{cases} \tag{4}$$

Hence, there exists the *disease-free* equilibrium:

$$E_0 \equiv (1,0,0)$$

Following the same discussion as in [8], it can be seen that for $R_0 < 1$, model (1)-(3) may admit none, one, or two endemic equilibria. For $R_0 > 1$, there exists a unique endemic equilibrium E.
Stability switches are also possible, so that the stability/instability of the endemic equilibria depends on the fulfillment of additional conditions on the parameters. For example, in the case $R_0 > 1$, it may be shown that the determinant of the jacobian matrix is always positive, whereas the sign of the trace depends on the parameters. The following result has been stated. Set

$$p = \frac{\alpha\mu}{k}, \quad q = \frac{\sigma}{\mu}, \quad B = \frac{k}{\mu}, \quad B_0 = \frac{4(1+q)^2}{q^2(q+2)};$$

$$p_0 = \frac{(4q+4+2\sqrt{4q^2+8q+4+B^2q^2})(q+1)}{2B^2q},$$

$$p_2 = \frac{B+2q+4+2\sqrt{(B+q+2)(q+2)}}{B^2}.$$

Proposition 1 ([8]). *If $B \leq B_0$ and $p < p_0$, or $B > B_0$ and $p < p_2$, then there exists an endemic equilibrium E, which is locally stable.*

The existence of the invariant manifold allows to refer to the reduced binary ODE:

$$\begin{cases} \dot{I} = kIS(1+\alpha I) - (\sigma+\mu)I \\ \dot{R} = \sigma I - (\mu+\gamma)R, \end{cases} \tag{5}$$

where, $S(R,I) = 1 - I - R$, and $(I(t), R(t)) \in \Sigma$, where,

$$\Sigma = \left\{(I,R) \in \mathbf{R}^2 : I, R \geq 0;\ I + R \leq 1\right\}.$$

3 Nonlinear stability/instability

Let us consider the perturbations to the generic equilibrium $u = S - S^*$, $v = R - R^*$ and $w = I - I^*$, we get from (4) and (5) the perturbation system:

$$\begin{cases} \dot{w} = (a - (\sigma+\mu))w + bv + \Phi(w,v) \\ \dot{v} = \sigma w - (\mu+\gamma)v, \end{cases} \tag{6}$$

where,

$$\Phi(w,v) = -(k + 2\alpha k I^*)vw + (\alpha k S^* - k - 2\alpha k I^*)w^2 - \alpha k v w^2 - \alpha k w^3, \tag{7}$$

and:

$$a = kS^*(1+2\alpha I^*) - kI^*(1+\alpha I^*), \quad b = -kI^*(1+\alpha I^*).$$

System (6) may be written:

$$\dot{\mathbf{v}} = J\mathbf{v} + \mathbf{G}(\mathbf{v}) \tag{8}$$

where: $\mathbf{v} \equiv (V_1, V_2) = (w, v)^T$; $J = [a_{ij}]$, where its entries are given by

$$a_{11} = a - (\sigma+\mu); \quad a_{12} = b; \quad a_{21} = \sigma; \quad a_{22} = -(\mu+\gamma),$$

and $\mathbf{G} = (\Phi, 0)^T$.
Denote the principal invariants of the jacobian matrix J as: $I = a_{11} + a_{22}$, and $A = a_{11}a_{22} - a_{12}a_{21}$. Set $V_1 = \alpha_0 v_1$, and $V_2 = \beta_0 v_2$, where α_0 and β_0 are positive constants to be suitably chosen later. Let be $\mu_0 = \alpha_0/\beta_0$. System (8) may be written:

$$\begin{cases} \dot{v}_1 = a_{11}v_1 + \mu_0^{-1}a_{12}v_2 + \alpha_0^{-1}g_1(\alpha_0 v_1, \beta_0 v_2) \\ \dot{v}_2 = \mu_0 a_{21}v_1 + a_{22}v_2. \end{cases} \tag{9}$$

We are in position to state the main result of this paper:

Theorem 1. *If $I < 0$, and $A > 0$, then the null solution of (8) is nonlinearly (locally) asymptotically exponentially stable. If $I > 0$ and $A > 0$, or $A < 0$, then it is unstable.*

Proof. We give only the sketch. For further details see [2], and [10], [11], [12], where a similar result has been obtained in the context of reaction diffusion systems. We begin with stability. Let $A > 0$ and $I < 0$, so that E is linearly stable. Introduce the following Lyapunov function:

$$V = \frac{1}{2}\left[A(v_1^2 + v_2^2) + (a_{11}v_2 - \mu_0 a_{21}v_1)^2 + (\mu_0^{-1}a_{12}v_2 - a_{22}v_1)^2\right] \tag{10}$$

Along the solutions of system (9) it turns out:

$$\dot{V}\Big|_{(9)} = AI(v_1^2 + v_2^2) + \Psi(\alpha_0 v_1, \beta_0 v_2), \tag{11}$$

where,

$$\Psi = \quad \alpha_0^{-1}\left(\alpha_1 v_1 - \alpha_3 v_2\right) g_1(\alpha_0 v_1, \beta_0 v_2),$$

and,

$$\alpha_1 = A + \mu_0^2 a_{21}^2 + a_{22}^2; \quad \alpha_3 = \mu_0 a_{11}a_{21} + \mu_0^{-1}a_{12}a_{22}. \tag{12}$$

It is easy to check that Φ is infinitesimal of higher order than $\sqrt{V_1^2 + V_2^2}$. This implies that there exist two positive constants k_0 and δ_0, such that:

$$|\Phi| \leq \delta_0 \left(\alpha_0^2 v_1^2 + v_2^2\right)^{k_0 + \frac{1}{2}}.$$

Now assume, for the sake of simplicity, that it is possible to choose α_0 and β_0 in such a way that $\alpha_3 = 0$ (for example, this is possible if $a_{11}a_{12}a_{21}a_{22} < 0$). Then we get the estimate: $\dot{V}\Big|_{(9)} \leq -\eta V$, for $V_0^k < d_1 d_2^{-1}$, where $\eta = d_1(1 - d_2 d_1^{-1} V_0^k)$, and d_1, d_2 are positive constants expressed in terms of the system parameters. Hence, the exponential decay follows: $V \leq V_0 e^{-\eta t}$.
As for the instability, the functional (10) allows a direct application of the instability theorem of Chetaev ([9], p.39).

Remark 1. *In terms of the trivial equilibrium E_0, the inequalities $I < 0$, $A > 0$, can be written:*

$$\begin{aligned} &(\sigma + \mu)(1 - R_0) > 0 \\ &(\sigma + \mu)(R_0 - 1) - (\mu + \gamma) < 0. \end{aligned}$$

¿From Theorem 1, we get the following result concerning the stability / instability of the trivial equilibrium E_0: if $R_0 < 1$, then E_0 is nonlinearly stable. If $R_0 > 1$, then E_0 is unstable.

4 Global nonlinear asymptotic stability

As mentioned in Section 1, in the paper [7] it has been shown that the SIRS model (1) admits an unique endemic equilibrium, which is globally asymptotically stable in $\mathbf{R}^2_+$, provided that $R_0 > 1$ and the incidence rate satisfies some sufficient conditions. Precisely, it is required that:

$$\begin{array}{ll} f(S,I^*) < f(S^*,I^*) & \text{for all } 0 < S < S^*; \\ f(S,I^*) > f(S^*,I^*) & \text{for all } S > S^*; \end{array}$$

and,

$$\begin{array}{ll} I/I^* \leq f(S,I)/f(S,I^*) \leq 1 & \text{for } 0 < I \leq I^*; \\ 1 \leq f(S,I)/f(S,I^*) \leq I/I^* & \text{for } I \geq I^*. \end{array} \tag{13}$$

It can be seen that inequalities (13) rule out several incidence rates, as the ones considered, e.g. in [14], [15], which are convex (i.e. concave up) respect to the variable I.

In this paper, the use of the Lyapunov Direct Method, through the peculiar Lyapunov function (10), allows to show that the conditions for the linear stability are necessary and sufficient for the local nonlinear stability of hyperbolic equilibria, according to Theorem 1. This result imply no restrictions on the incidence rate about its monotonicity and concavity with respect to the infective class. Furthermore, if one is able to 'control' the functional Ψ in (11), then sufficient conditions for the global stability may be obtained. This is the main goal of this Section.

We know that the solution to (1) are bounded, so that we can consider uniformly bounded perturbations. That is, we assume that a positive constant m there exist, such that:

$$|v_1| < m, \; |v_2| < m.$$

This observation, taking into account of (7), allows to get the estimate:

$$\begin{array}{rl} \Psi \leq & m\alpha_1\alpha_0^{-1}\left[|k+2\alpha kI^*|\,(\alpha_0^2+1)(v_1^2+v_2^2)+\right. \\ & \left.+\,|\alpha kS^*-k-2\alpha kI^*|\,\alpha_0^2 v_1^2 + 2\alpha k m\alpha_0^2 v_1^2\right], \end{array}$$

that is,

$$\Psi \leq \delta_1(v_1^2+v_2^2),$$

where

$$\begin{array}{rl} \delta_1 = & m\alpha_1\alpha_0^{-1}\left[2\,|k+2\alpha kI^*|\,(\alpha_0^2+1)+\right. \\ & \left.+\,|\alpha kS^*-k-2\alpha kI^*|\,\alpha_0^2 + 2\alpha k m\alpha_0^2\right]. \end{array}$$

Hence, it follows:

$$\dot{V}\Big|_{(9)} \leq (-d_3+d_4)V,$$

where,

$$d_3 = \frac{A\,|I|}{k_2}; \quad d_4 = \frac{\delta_1}{k_1}. \tag{14}$$

and k_1 and k_2 are:

$$k_1 = \frac{A}{2}; \quad k_2 = \max\left\{A, 2(a_{11}^2+\alpha_0^2 a_{21}^2), 2\left(\frac{a_{12}^2}{\alpha_0^2}+a_{22}^2\right)\right\}.$$

Hence:

$$V \leq V(0)e^{-(d_3-d_4)t}.$$

We can summarize the result of this section in the following theorem:

Theorem 2. *When the SIRS epidemic model (1) with the incidence force given by (3) admits a locally stable equilibrium, then such equilibrium is nonlinearly globally stable, provided that* $d_3 > d_4$*, where* d_3 *and* d_4 *are given by (14).*

5 Conclusions

In this paper we provide a direct application to epidemic models of a recent approach to nonlinear stability, due to S.R. We consider a SIRS epidemic model with convex incidence rate. This case is ruled out by the sufficient conditions for global stability obtained in [7] through the Lyapunov Direct Method. Here, we still use such method and, by considering a peculiar Lyapunov function, we are able to find exact coincidence between the linear and local nonlinear stability conditions for the equilibria. We further find sufficient conditions for the global stability of the endemic equilibrium. Hence, we enlarge (with respect to what established in [7]) the class of incidence rates ensuring the global nonlinear stability of the endemic equilibria. Further future investigations will refer to epidemic models with general nonlinear incidence rates, and will be the subject of a forthcoming paper, [2].

Acknowledgements

The present work has been performed under the auspices of the italian National Group for Mathematical Physics (GNFM-Indam) *Waves and Stability in Continuous Media* and the granted scientific project entitled *Analisi di due metodi per la stabilità ed applicazioni a modelli ODE e PDE in dinamica delle popolazioni.*

References

[1] R.M. Anderson, R.M. May: *Infectious diseases in humans: dynamics and control.* Oxford University Press, Oxford (1991).

[2] B. Buonomo, S. Rionero: On the Lyapunov stability for SIRS epidemic models with general incidence rate. In progress.

[3] S.N. Busenberg, K. Cooke: *Vertically trasmitted desease.* Springer, Berlin (1993).

[4] V. Capasso: *Mathematical Structures of Epidemic Systems.* Lecture Notes in Biomath., 97. Springer-Verlag, Berlin (1993).

[5] H.W. Hethcote, The mathematics of infectuous diseases. SIAM Rev. **42**, n.4, 599-653 (2000).

[6] A. Korobeinikov: Lyapunov functions and global stability for SIR and SIRS epidemiological models with non-linear transmission. Bull. Math. Biol., **30**, 615-626 (2006).

[7] A. Korobeinikov: Global properties of infectuous disease models with nonlinear incidence. Bull. Math. Biol., **69**, 1871-1886 (2007).

[8] Y. Jin, W. Wang, S. Xiao: An SIRS model with a nonlinear incidence rate. Chaos, Solit. Fract., **34**, 1482-1497 (2007).

[9] J. La Salle, S. Lefschetz: *Stability by the Lyapunov's Direct Method*, Academic Press, New York (1961).

[10] S. Rionero: L^2 stability of solutions to a nonlinear binary reaction-diffusion system of P.D.E.s. Rend. Mat. Acc. Lincei, s.9, Vol. 16, 227-238 (2005).

[11] S. Rionero: A nonlinear L^2 stability analysis for two species dynamics with dispersal. Math. Biosci. Engineering, **3**, n.1, 189-204 (2006).

[12] S. Rionero: A rigorous reduction f the L^2-stability of the solutions to a nonlinear binary reaction-diffusion system of P.D.E.s to the stability of the solutions to a linear binary system of ODE's. J. Math. Anal. Appl., **319**, 377-397 (2006).

[13] S. Rionero: Nonlinear L^2-stability analysis for two-species population dynamics in spatial ecology under Neumann boundary data. Rend. Circ. Mat. Palermo. Serie II, Suppl. **78**, 273-283 (2006).

[14] P. van den Driessche, J. Watmough: A simple SIS epidemic model with backward bifurcation, J. Math. Biol., **40** 525-540 (2000).

[15] D. Xiao, S. Ruan: Global analysis of an epidemic model with nonmonotone incidence rate, Math. Biosci., **208** 419-129 (2007).

On the Dynamics of a Binary Model for the Competition between Bacteria and Innate Immune System

M. Cerasuolo, P. Fergola, S. Rionero

Dipartimento di Matematica e Applicazioni *R. Caccioppoli*,
Università degli Studi di Napoli *Federico II*,
via Cintia, 80126 Napoli, Italy.

Abstract

A competition between bacteria and innate immune system of invertebrates is described by means of a pair of nonlinear reaction-diffusion PDE's, with constant coefficients. In the system it is also considered the presence of the Quorum Sensing (Q.S.) phenomenon. The qualitative properties of the associated kinetic system have been studied. The local stability properties have been proved for some biological meaningful solutions. Some results on the linear stability properties of the solutions of the PDE model with diffusion have also been given.

Keywords: invertebrate species, innate immunity, bacterial infections, quorum sensing, reaction-diffusion system

1 Introduction

The immune system of invertebrate animals, which lack adaptive immune systems, only consists of the innate response which reacts to the presence of pathogens (bacteria, virus, fungi) (see [1]). The major innate immune systems include receptor cells, which detect invaders, and phagocytic cells, such as macrophage-like, neutrophil-like, or dendritic cells, which ultimately engulf all the *non-self* cells and then internalize, process, and kill them. Therefore, in this sense, we can describe bacterial infections in the invertebrate species as a *competition* between bacteria and their immune system. The innate response is non-specific because the strategy is irrespective of the bacterial species. The immune system can suffice to clear the pathogen in most cases, but sometimes it is insufficient. Bacteria can indeed be able to overcome the innate response and successfully colonize and infect the host. By assuming in our mathematical modeling approach a macroscopic point of view, we will give a qualitative description of such a competition by means of a pair of nonlinear reaction-diffusion PDE's.

$$\begin{cases} \dfrac{\partial B}{\partial t} = aB + \gamma_1 \Delta B - \left(\dfrac{a}{k}B + \dfrac{eI-\gamma}{k_1+B}\right)B \\ \\ \dfrac{\partial I}{\partial t} = s - cI + dB + \gamma_2 \Delta I \end{cases} \quad \text{in}\, \Omega \times \mathbb{R}^+ \tag{1}$$

under the Neumann boundary conditions

$$\begin{cases} \nabla B \cdot \underline{n} = 0 \\ \\ \nabla I \cdot \underline{n} = 0 \end{cases} \quad \text{on}\, \partial\Omega \times \mathbb{R}^+ \tag{2}$$

where $\Omega \subset \mathbb{R}^3$ is a bounded smooth domain, $\underline{n}$ is the unit outer normal vector on $\partial\Omega$. $B(t)$ and $I_R(t)$ denote the concentration of the bacteria and of the innate immune system cells at time t, respectively.

Innate response has a source term s and a half-life term c, the source term includes a wide range of cells involved in the first wave line of defence of the host (such as natural killer cells polymorphonuclear cells, macrophages and dendritic cells). The response is enhanced and sustained by signals that have been captured by bacteria load.

The bacteria population has a net growth term, represented by a logistic function $aB(1-\frac{B}{k})$. The presence of Quorum Sensing [2] is qualitatively represented by means of the further growth term $\frac{\gamma B}{k_1+B}$ where the constant k_1 represents a given concentration of bacteria beyond which the bacterial density tends to saturation. Finally, we assume that the bacterial population is cleared by innate immunity and this interaction between bacteria and immune system cells is represented by an Holling-type II term $(\frac{eB}{k_1+B}I)$, where the same constant k_1 represents the concentration of bacteria beyond which the attack capacity of the innate system tends to saturation. All the parameters are positive constants.

The plane of the paper is as follows. Section 2 contains some necessary preliminaries. In Section 3 we study the associated kinematic system whose biological meaningful equilibria represent infected or uninfected steady-states. In Section 4 sufficient conditions for linear stability are found for the P.D.E binary system by using a peculiar Liapunov functional recently introduced (see, for instance, [3]) in the context of reaction-diffusion problems.

2 Preliminaries

We assume that $\Omega \subset \Re^3$ be an open bounded set having the interior cone property [4] with a $C^{2+\eta}$ boundary $\partial\Omega$ {[5], pg.626}. We remark that:

i) the perturbations equations are given by

$$\begin{cases} \dfrac{\partial U}{\partial t} = a_1 U - a_2 V + \gamma_1 \Delta U + f \\ \\ \dfrac{\partial V}{\partial t} = dU - cV + \gamma_2 \Delta V \end{cases} \tag{3}$$

$$\begin{cases} \nabla U \cdot \underline{n} = 0 \\ \\ \nabla V \cdot \underline{n} = 0 \end{cases} \quad \text{on}\, \partial\Omega \times \mathbb{R}^+ \tag{4}$$

with

$$\begin{cases} f = \left[\dfrac{(eI^* - \gamma)k_1}{(k_1 + B^*)(k_1 + B^* + U)} - \dfrac{a}{k}\right] U^2 - \dfrac{ek_1}{(k_1 + B^*)(k_1 + B^* + U)} UV, \\ a_1 = \dfrac{a(k - 2B^*)(k_1 + B^*)^2 - kk_1(eI^* - \gamma)}{k(k_1 + B^*)^2}, \qquad a_2 = \dfrac{eB^*}{k_1 + B^*}, \\ U = B - B^*, \qquad V = I - I^*. \end{cases} \tag{5}$$

ii) Denoting by $\|\cdot\|$, $\langle\cdot,\cdot\rangle$, the $L^2(\Omega)$ norm and its scalar product and setting (α, β positive scalings to be chosen later)

$$\begin{cases} U = \alpha u, & V = \beta v, \quad \mu = \frac{\alpha}{\beta}, \\ a_{11} = a_1 - \gamma_1 \overline{\alpha}, & a_{12} = -a_2, \\ a_{21} = d, & a_{22} = -(c + \gamma_2 \overline{\alpha}), \\ 0 < \overline{\alpha} = \min\limits_{W^{1,2}(\Omega)} \dfrac{\|\nabla\varphi\|^2}{\|\varphi\|^2}, & \nabla\varphi \neq 0. \end{cases} \tag{6}$$

$W^{1,2}(\Omega)$ being the space of the functions φ verifying (10) (space of the kinematic admissible perturbations) one obtain

$$\begin{cases} \dfrac{\partial u}{dt} = a_{11}u + \dfrac{1}{\mu}a_{12}v + f^* + \dfrac{1}{\alpha}f(U,V) \\ \dfrac{\partial v}{dt} = \mu a_{21}u + a_{22}v + g^* \end{cases} \tag{7}$$

under the boundary conditions

$$\begin{cases} \nabla u \cdot \underline{n} = 0 \\ \nabla v \cdot \underline{n} = 0 \end{cases} \quad \text{on } \partial\Omega \times \mathbb{R}^+ \tag{8}$$

and, with

$$\begin{cases} f^* = \gamma_1(\Delta u + \overline{\alpha}u) \\ g^* = \gamma_2(\Delta v + \overline{\alpha}v) \end{cases} \tag{9}$$

iii) $\overline{\alpha}$ is the lowest positive eigenvalue of the spectral problem

$$\begin{cases} \nabla\varphi \neq 0 \\ \Delta\varphi + \lambda\varphi = 0 & \text{in } \Omega \\ \nabla\varphi \cdot \underline{n} = 0 & \text{in } \partial\Omega \end{cases} \tag{10}$$

3 The associated kinetic system

In this Section we analyze and discuss the main qualitative properties of the O.D.E. system obtained by (1). Moreover, we calculate its steady state solutions. Let us rewrite the equations $(1)_1$, $(1)_2$ as follows:

$$\begin{cases} \dfrac{dB}{dt} = aB\left(1 - \dfrac{B}{k}\right) - \dfrac{eIB}{k_1 + B} + \dfrac{\gamma B}{k_1 + B} \\ \dfrac{dI}{dt} = s - cI + dB \end{cases} \tag{11}$$

For biological relevance, we consider non-negative initial conditions.

Lemma 1. *Any solution of (11) with $(B(0), I(0)) \in \mathbb{R}^2_+$ remains positive whenever it exists.*

Proof. Consider the first equation in (11)

$$\frac{dB(t)}{dt} = B(t)\left[a\left(1 - \frac{B}{k}\right) - \frac{eI - \gamma}{k_1 + B}\right]$$

with $B(0) \geq 0$. Then

$$B(t) = B(0)\exp\left\{\int_0^t \left[a\left(1 - \frac{B(r)}{k}\right) - \frac{eI(r) - \gamma}{k_1 + B(r)}\right] dr\right\} > 0, \ t \geq 0. \qquad (12)$$

Of course, if $B(0) = 0$, $B(t) = 0 \ \forall t \geq 0$. Let us consider now the equation $(11)_2$, we have

$$\frac{dI}{dt} \geq s - cI, \qquad I(0) \geq 0$$

hence

$$I(t) \geq I(0)e^{-ct} + \frac{s}{c}(1 - e^{-ct}) > 0, \ t \geq 0. \qquad (13)$$

This completes the proof of the positivity.

Lemma 2. *Any solution of (11) is bounded.*

Proof. Because of positivity of solutions, the first equation in (11) give

$$\frac{dB(t)}{dt} < aB(t)\left(1 - \frac{B}{k}\right) + \gamma$$

i.e.

$$\limsup_{t \to +\infty} (B(t)) < \overline{B}$$

with $\overline{B} = k + \frac{\gamma}{2a}$. Therefore, there exists a sufficiently small ϵ such that, $B(t) < \overline{B} + \epsilon$ and the last equation gives

$$\frac{dI}{dt} < s + d(\overline{B} + \epsilon) - cI$$

thus implying that (by letting $\epsilon \to 0$)

$$\limsup_{t \to +\infty} (I(t)) < \frac{s + d\overline{B}}{c}.$$

This prove the boundedness.

Lemma 3. *If* $k_1 > \dfrac{e(2adk + 2as + d\gamma)}{2a^2c}$ *then system (11) is permanent.*

Proof. We say that system (11) is permanent (or uniformly persist) if there exist positive constants, m and M ($m < M$), independent of initial conditions and such that for the solutions of (11) we have:

$$\begin{aligned} &\max\{\limsup_{t\to+\infty} B(t), \limsup_{t\to+\infty} I(t)\} \le M \\ &\min\{\liminf_{t\to+\infty} B(t), \liminf_{t\to+\infty} I(t)\} \ge m. \end{aligned} \tag{14}$$

Let us first consider the limsup. From the positivity we have

$$\left\{\limsup_{t\to+\infty} B(t), \limsup_{t\to+\infty} I(t)\right\} \le (\overline{B}, \overline{I})$$

where $\overline{B} = k + \dfrac{\gamma}{2a}$, $\overline{I} = \dfrac{s + d\overline{B}}{c}$. If we choose

$$M = \max\{\overline{B}, \overline{I}\},$$

then there exists $M > 0$ such that the first inequality of (14) holds true.
Consider now the liminf. From (13) we have

$$\liminf_{t\to+\infty} I(t) \ge \frac{s}{c} := \underline{I}. \tag{15}$$

From (14) for sufficiently large $t > 0$ and small $\epsilon > 0$, we have

$$\frac{dB(t)}{dt} \ge aB(t)\left(1 - \frac{B}{k}\right) - \frac{(e\overline{I} + \epsilon)}{k_1} B(t) = aB(t)\left(1 - \frac{(e\overline{I} + \epsilon)}{ak_1} - \frac{B(t)}{k}\right).$$

Hence getting $\epsilon \to 0$ we have that if

$$ak_1 > e\frac{s + d(k + \frac{\gamma}{2a})}{c} \ (= e\overline{I})$$

then

$$\liminf_{t\to+\infty} B(t) \ge \frac{ak_1 - e\overline{I}}{ak_1} k := \underline{B}. \tag{16}$$

Hence, in the hypotheses of the lemma we have

$$\left\{\liminf_{t\to+\infty} B(t), \liminf_{t\to+\infty} I(t)\right\} \ge (\underline{B}, \underline{I}) \tag{17}$$

where the constants on the right side of (17) are positive. Then, if we choose

$$m = \min\{\underline{B}, \underline{I}\},$$

then there exists $m > 0$ such that the second inequality of (14) holds true.

Moreover, about the existence and stability properties of the steady state solutions, by using the same line of proof used in [6], we obtain the following Theorems:

Theorem 1. *System (11) admits*

i. *For all values of parameters there exists the uninfected equilibrium* $E_0 = \left(0, \frac{s}{c}\right)$.

ii. *If* $k_1 > \dfrac{se - \gamma c}{ac}$ *then there exists one infected equilibrium* $E_1^* = \left(B_1^*, \dfrac{s + dB_1^*}{c}\right)$, *whose B-component is given by*

$$B_1^* = \frac{-ked - k_1 ac + akc + \sqrt{4akc(-se + k_1 ac + \gamma c) + (ked + ak_1 c - akc)^2}}{2ac}.$$

iii. *If* $e > \dfrac{\gamma c}{s}$ *and* $\dfrac{-k(ac + ed) + 2\sqrt{ack(se + ekd - \gamma c)}}{ac} < k_1 < \dfrac{se - \gamma c}{ac}$ *then there exist two infected equilibria* $E_i^* = \left(B_i^*, \dfrac{s + dB_i^*}{c}\right)$, $i = 1, 2$ *whose B-components are given by*

$$B_i^* = \frac{-ked - k_1 ac + akc \pm \sqrt{4akc(-se + k_1 ac + \gamma c) + (ked + ak_1 c - akc)^2}}{2ac}.$$

Theorem 2. *If* $e > \dfrac{\gamma c}{s}$ *then the equilibrium* E_0 *is linearly stable if and only if* $k_1 < \dfrac{es - \gamma c}{ac}$ *otherwise it is unstable.*

Proof. Let us consider system (11) perturbed around the E_0 solution. Therefore, by setting

$$\begin{cases} u = B \\ v = I - \dfrac{s}{c} \end{cases} \tag{18}$$

(11) can be rewritten as follows

$$\begin{cases} \dfrac{du}{dt} = \left(a + \dfrac{\gamma}{k_1} - \dfrac{es}{ck_1}\right) - \left(\dfrac{au}{k} + \dfrac{\gamma u + k_1 ev - \frac{es}{c}}{k_1(k_1 + u)}\right) u^2 \\ \dfrac{dv}{dt} = -cv + du \end{cases} \tag{19}$$

In this way we obtain

$$\begin{bmatrix} a + \dfrac{\gamma}{k_1} - \dfrac{es}{ck_1} & 0 \\ d & -c \end{bmatrix}$$

from which

$$I = a + \frac{\gamma}{k_1} - \frac{es}{ck_1} - c$$

$$A = -c\left(a + \frac{\gamma}{k_1} - \frac{es}{ck_1}\right)$$

Because of the hypotheses we have $a + \frac{\gamma}{k_1} - \frac{es}{ck_1} < 0$, so

$$I < 0 \quad \text{and} \quad A > 0.$$

Therefore the equilibrium E_0 is locally asymptotically stable.

Theorem 3. *About the internal equilibria it turns out that:*

1. *If* $c \geq a$ *or* $c < a$ *and* $k_1 > k\left[1 + \frac{c}{a}\left(3 - 2\sqrt{2\left(\frac{a}{c}+1\right)}\right)\right]$ *then the equilibrium* E_1^**, if it exists, it is linearly stable.*
2. *If the equilibrium* E_2^* *exists then it is unstable.*

Proof. Let us consider system (11) perturbed around the E_i^* solution. By setting

$$\begin{cases} u = B - B_i^* \\ v = I - I_i^* \end{cases} \tag{20}$$

we have that (11) can be rewritten as follows:

$$\begin{cases} \dfrac{du}{dt} = & \dfrac{aB_i^*(k - k_1 - 2B_i^*)u}{k(k_1 + B_i^*)} - \dfrac{eB_i^*}{k_1 + B_i^*}v + \\ & -\dfrac{a(B^{*2} + B_i^* u + 3B_i^* k_1 - kk_1 + uk_1 + k_1^2)}{k(k_1 + B_i^*)(k_1 + B_i^* + u)}u^2 - \dfrac{ek_1}{(k_1 + B_i^*)(k_1 + B_i^* + u)}uv \\ \dfrac{dv}{dt} = & -cv + du \end{cases} \tag{21}$$

In this way we obtain

$$\begin{bmatrix} \frac{aB_i^*(k-k_1-2B_i^*)}{k(k_1+B_i^*)} & -\frac{eB_i^*}{k_1+B_i^*} \\ d & -c \end{bmatrix}$$

from which

$$I(E_i^*) = \frac{aB_i^*(k - k_1 - 2B_i^*)}{k(k_1 + B_i^*)} - c = \frac{-2aB_i^{*2} + (ak - ck - ak_1)B_i^* - ckk_1}{k(k_1 + B_i^*)}$$

$$A(E_i^*) = -c\left(\frac{aB_i^*(k - k_1 - 2B_i^*)}{k(k_1 + B_i^*)}\right) + \frac{edB_i^*}{k_1 + B_i^*} = \frac{B_i^*(2acB_i^* - ack + dek + ack_1)}{k(k_1 + B_i^*)} \qquad i = 1, 2$$

1. In the hypotheses of the theorem we have that

$$A(E_1^*) = \sqrt{(-ack + dek + ack_1)^2 + 4ack(-es + c\gamma + ack_1)} > 0$$

   and

$$I(E_1^*) < 0$$

   therefore E_1^* is locally asymptotically stable.

2. From the expression of B_2^* it follows that $A(E_2^*) < 0$ for all parameter values, therefore E_2^* is unstable.

4 Linear Stability and Instability results for the diffusion problem (1)-(2)

In this section, for sake of brevity, we will limit ourselves to give, without proofs, some results concerning the stability properties of the problem with diffusion (1)-(2) with respect to the perturbations

$$\begin{cases} u = \sum_{n=1}^{\infty} X_n(t)\varphi_n(x) \\ v = \sum_{n=1}^{\infty} Y_n(t)\varphi_n(x) \end{cases} \tag{22}$$

with (22) derivable, term by term, once with respect to t and twice with respect to the spatial variables.

Theorem 4. *Let*

$$a_1 < \gamma_1 \overline{\alpha} \tag{23}$$

then the equilibrium E^*, $(i = 1, 2)$ *is linearly stable.*

Remark 1. *We can observe that if (23) holds (where* a_1 *is computed with respect to the equilibrium* E_2^**) then* E_2^**, unstable in the absence of diffusion, becomes stable and hence, in this case, we can assert that the diffusion has a stabilizing effect.*

Theorem 5. *Let* $a_1 > \gamma_1 \overline{\alpha}$ *and*

$$a_1 < \overline{a} = \min\{c; \overline{\alpha}\gamma_1 + \frac{da_{12}}{\gamma_2 \overline{\alpha} + c};, 2\overline{\alpha}\gamma_1 + \frac{c\gamma_1}{\gamma_2}\} \tag{24}$$

then the equilibrium E^* $(i = 1, 2)$ *is linearly stable.*

Theorem 6. *Let*

$$a_1 = \gamma_1 \overline{\alpha} \tag{25}$$

then the equilibrium E^* *is linearly stable.*

Theorem 7. *Let*

$$\begin{cases} c \geq a \vee c < a \wedge k_1 > k \left[1 + \frac{c}{a}\left(3 - 2\sqrt{2\left(\frac{a}{c} + 1\right)}\right)\right] \\ \gamma_1 < \min\left\{\frac{a_1\gamma_2}{c}; \frac{a_1}{\overline{\alpha}}\right\} \\ \gamma_2 > \dfrac{a_2 d - a_1 c - \gamma_1 c \overline{\alpha}}{\overline{\alpha}(a_1 - \gamma_1 \overline{\alpha})} \end{cases} \tag{26}$$

then the equilibrium E_1^* *stable in absence of diffusion, becomes unstable.*

Remark 2. *The hypotheses of Theorem 7 can be written equivalently as follows*

$$a_1 > \max\left\{\frac{\gamma_1 c}{\gamma_2}, \gamma_1 \overline{\alpha} + \frac{da_{12}}{\gamma_2 \overline{\alpha} + c}\right\}.$$

Remark 3. *For the proofs of Theorems 4 and 7 we refer to [7] where the longtime behaviour of the solutions is studied through the nonlinear stability of the steady states and the determination of the absorbing sets.*

Acknowledgement

This work has been performed under the auspicious of the GNFM-INDAM, PRIN *Waves and Stability in Continuous Media.* M. Cerasuolo also thanks the Postdoctoral fellow *Sull'Influenza dei Fenomeni Diffusivi nelle Infezioni Batteriche*, University of Naples Federico II, Department of Mathematics and Applications *R. Caccioppoli.*

References

[1] S. Iwanaga and B. L. Lee. Recent advances in the innate immunity of invertebrate animals. *Journal of Biochemistry and Molecular Biology*, 38(2):128–150, 2005.

[2] B.L. Bassler. How bacteria talk to each other: regulation of gene expression by quorum sensing. *Current Opinion in Microbiology*, 2:582, 1999.

[3] S. Rionero. A rigorous reduction of the L^2-stability of the solutions to a nonlinear binary reaction-diffusion system of p.d.es. to the stability of the solutions to a linear binary system of o.d.es. *Journal of Math. Anal. Appl.*, 319:377–397, 2006.

[4] R.A. Adams. *Sobolev Spaces.* Academic Press, New York, 1st edition, 1975.

[5] L.C. Evans. *Partial Differential Equations*, volume 19. AMS, Berkeley, 1997.

[6] P. Fergola, Juan Zhang, M. Cerasuolo, and Zhien Ma. On the influence of quorum sensing in the competition between bacteria and immune system of invertebrates. *in press*, 2008.

[7] M. Cerasuolo, P. Fergola, and S. Rionero. Influence of diffusion on the competition between bacteria and innate immune system of invertebrates. *to appear.*

Spatial Evolution in Linear Thermoelasticity

S. Chirita[a], *M. Ciarletta*[b]

[a]Faculty of Mathematics, Al. I. Cuza University of Iaşi,
Blvd. Carol I, nr. 11, 700506 Iaşi, Romania

[b]DIIMA, University of Salerno,
Via Ponte Don Melillo, 84084 Fisciano, Italy

Abstract

In the present paper we consider a prismatic cylinder occupied by an anisotropic homogeneous compressible linear thermoelastic material that is subject to zero body force and heat supply and zero displacement–temperature variation on the lateral boundary. The elasticity tensor is strongly elliptic and the conductivity tensor is positive. The motion is induced by a harmonic time–dependent displacement–temperature variation specified pointwise over the base. We establish some spatial estimates for appropriate cross–sectional measures associated with the amplitude of the corresponding harmonic vibrations that describe how the amplitude evolves with respect to the axial distance to the excited base. In fact, in the low frequency range we establish some *exponential* estimates describing the spatial evolution of the amplitude of vibration which confirms the observations made by Boley in related contexts.

1 Introduction

One of the important problems in solid and structural mechanics is to determine the extent of mechanical and thermal local or edge effects in structural elements. It is well known that the global structure response to external stimuli depends heavily on the motions of its inner structure. In fact, in the design and construction of buildings, bridges, aircraft, nuclear reactors and automobiles, for example, it is important to establish the depth to which local stresses, such as those produced by fasteners and at joints, or local heat flux, can penetrate girders, I–beams, braces, and other similar structural elements. Thus, the determination of the effects of local stresses in a structure allows the engineer to make a distinction between the global structure, where strength of materials approximations may be used and local load–bearing portions which require a separate, more elaborate analysis based on some exact theory of solid mechanics (for example, elasticity and thermoelasticity theories). In order to achieve this in the statics of structures, the standard procedure used in engineering practice relies on the Saint–Venant principle, a concept involving load resultants. Thus Saint–Venant's principle is appealed to in neglecting local or end effects and experience with homogeneous isotropic materials in linear elasticity has served to establish this standard procedure. It is a fundamental principle also forming the basis for static mechanical testing of materials.

Thus property measurements are made in a suitable gage section where uniform stress and strain states are induced and local effects due to clamping of the specimen are neglected on invoking Saint–Venant's principle. Barré de Saint–Venant in his fundamental memoirs [1,2] formulated what we now call Saint–Venant's problem concerned with a linearly elastic cylinder in equilibrium under the action of loads that arise solely from tractions applied to its plane ends. Saint–Venant conjectured that the detailed mode of application and distribution of end loads should be of less significance than their respective resultant forces and moments. Thus arose the subsequent assertions in the literature, which came to be known as Saint–Venant's principle, regarding the effect on stresses in an elastic body caused by replacing tractions over a portion of its surface with statically equivalent ones. In the framework of the linear theory of elasticity, the problem of comparing two statically equivalent systems of traction is reduced, by applying superposition, to the study of the stresses in the same body subject to arbitrary self–equilibrated tractions on part of the boundary with the remainder of the boundary being traction–free. The simplifications arising in elasticity theory from consideration of resultant boundary conditions instead of mathematically exact pointwise conditions have been the key to widespread application of the subject. Thus, theories for strength of materials, plates, and shells rely on such relaxed boundary conditions for their development.

Issues related to those just discussed have been considered for transient solutions. The feasibility of a Saint–Venant principle in this context was extensively discussed in literature on subject. Boley [3,4] was apparently the first to address the question whether principles of the Saint–Venant type apply to the transient heat conduction. A complete discussion on this argument as well as for elastodynamics can be found in review articles by Horgan and Knowles [5] and Horgan [6,7].

As regards steady state solutions, Flavin and Knops [8] have carried out an analysis of spatial decay for certain damped acoustic and elastodynamic problems in the low frequency range which substantiates the early work of Boley. Extension of these results to nonlinear elastodynamics and linear anisotropic elastodynamics are described by Flavin et al. [9] and Knops [10]. Further contributions in linear thermoelastodynamics are given by Chiriţă [11, 12], while in linear micropolar elasticity these are given by Aron and Chiriţă [13]. It should be noted that all of the harmonic vibrations investigations mentioned in the foregoing were concerned with appropriate low frequency range and they were established under the positive definiteness assumptions on the elasticity coefficients.

On the other hand, for linear elastodynamics it was outlined in [5-7] that high–frequency effects may propagate with little spatial attenuation. So one would not expect to find unqualified decay estimates of the kind discussed here in problems involving elastic wave propagation, even if the end loads are self–equilibrated at each instant.

In the present paper we consider a prismatic cylinder occupied by an anisotropic linear thermoelastic material and subject to zero body force and zero heat supply and zero lateral boundary conditions and zero initial conditions. The motion is induced by a harmonic time–dependent displacement–temperature variation specified pointwise over the base. The elasticity tensor is strongly elliptic and the conductivity tensor is positive.

The primary purpose of the present paper is to examine how the amplitude of the harmonic vibration evolves with respect to the axial variable. To this end we associate with the amplitude of the harmonic vibration in concern, an appropriate cross–sectional integral function and further we prove that the strong ellipticity of the elasticity tensor and the positiveness of the conductivity tensor assure that it is an acceptable measure. Then we establish some differential inequalities that describe the spatial behavior of the corresponding measure. In fact, in the low frequency range we establish some exponential estimates describing the spatial evolution of the amplitude of vibration, which confirms the observations made by Boley in related contexts. While in the high frequency range we expect certain algebraical estimates that predict a decay or growth slower than that for low frequency range. The results are illustrated for transversely isotropic materials.

2 Formulation of problem

Consider a prismatic cylinder $B \subset \mathbb{R}^3$ whose bounded uniform cross–section $D \subset \mathbb{R}^2$ has piecewise continuously differentiable boundary ∂D. The origin of a rectangular Cartesian coordinate system is located in the cylinder's base and the positive x_3–axis is directed along that of the cylinder. It is convenient to introduce the further abbreviation

$$B_z = \{x \in B : z > x_3\} \tag{2.1}$$

and, moreover, we employ $D(x_3)$ to indicate that relevant quantities are to be evaluated over the cross–section whose distance from the origin is x_3.

The cylinder is occupied by an anisotropic compressible thermoelastic material and is subject to a deformation in which the displacement–temperature variation field $(u_r, \theta)(x,t)$ is a smooth function satisfying the requirements of the classical dynamical theory of thermoelasticity [14]. The corresponding stress tensor $\mathbf{S}(x,t)$ has Cartesian components given by

$$S_{rs} = C_{rsmn} u_{m,n} + M_{rs}\theta, \tag{2.2}$$

where the constant elasticities C_{rsmn} and stress–temperature components M_{rs} possess the symmetries

$$C_{rsmn} = C_{mnrs} = C_{srmn}, \quad \mathrm{M_{rs}} = \mathrm{M_{sr}}, \tag{2.3}$$

and the elasticities satisfy the strong ellipticity condition

$$C_{rskl} m_r m_k n_s n_l > 0 \ \ \mathrm{forallnon--zerovectors} \left(\mathrm{m_1, m_2, m_3}\right), \left(\mathrm{n_1, n_2, n_3}\right). \tag{2.4}$$

The heat flux vector $\mathbf{Q}(x,t)$ has Cartesian components

$$Q_r = -K_{rs}\theta_{,s}, \tag{2.5}$$

where K_{rs} are the constant components of the conductivity tensor and they satisfy

$$K_{rs}\theta_{,r}\theta_{,s} \geq 0. \tag{2.6}$$

The cylinder is set in motion subject to a pointwise prescribed base harmonic time–dependent displacement–temperature variation, zero body–force and heat supply and zero displacement–temperature variation on the lateral surface. Furthermore, the prescribed displacement–temperature variation is such that a classical solution exists on the interval $[0,\infty)$. Consequently, the problem to be considered is specified by

$$(C_{rskl}u_{k,l}+M_{rs}\theta)_{,r}=\rho\ddot{u}_s,\quad (\mathrm{x},\mathrm{t})\in \mathrm{B}\times[0,\infty), \tag{2.7}$$

$$(K_{rs}\theta_{,s})_{,r}+\theta_0 M_{rs}\dot{u}_{r,s}=c\dot{\theta},\quad (\mathrm{x},\mathrm{t})\in \mathrm{B}\times[0,\infty),$$

$$u_r(x,0)=u_r^0(x),\quad \theta(\mathrm{x},0)=\theta^0(\mathrm{x}),\quad \dot{\mathrm{u}}_\mathrm{r}(\mathrm{x},0)=\dot{\mathrm{u}}_\mathrm{r}^0(\mathrm{x}),\quad \mathrm{x}\in\mathrm{B}, \tag{2.8}$$

$$u_r(x,t)=0,\ \theta(\mathrm{x},\mathrm{t})=0,\quad (\mathrm{x},\mathrm{t})\in\partial\mathrm{D}\times[0,\mathrm{L}]\times[0,\infty), \tag{2.9}$$

$$u_r(x,t)=f_r(x_1,x_2)\,e^{i\omega t},\quad \theta(\mathrm{x},\mathrm{t})=\mathrm{g}(\mathrm{x}_1,\mathrm{x}_2)\,\mathrm{e}^{\mathrm{i}\omega\mathrm{t}},\quad (\mathrm{x},\mathrm{t})\in\mathrm{D}(0)\times[0,\infty), \tag{2.10}$$

$$u_r(x,t)=0,\ \theta(\mathrm{x},\mathrm{t})=0,\quad (\mathrm{x},\mathrm{t})\in\mathrm{D}(\mathrm{L})\times[0,\infty), \tag{2.11}$$

in the case where L is finite (say). In the limiting case $L\to\infty$ a condition of the type (2.11) is unnecessary. In the above relations we have used a superposed dot for denoting differentiation with respect to time and a subscript comma indicates partial differentiation. Moreover, ρ is the constant positive mass density, c is the specific heat, θ_0 is the uniform reference temperature, ω is a positive constant, $u_r^0(x)$, $\theta^0(x)$, $\dot{u}_r^0(x)$, $f_r(x_1,x_2)$ and $g(x_1,x_2)$ are prescribed differentiable functions compatible with the initial and lateral boundary conditions and $i=\sqrt{-1}$ is the complex unit. We are interested in the study of the spatial behavior of the solution (u_r,θ) of the above initial boundary value problem (2.7) to (2.11).

To this end we use the following decomposition

$$u_r=U_r(x,t)+v_r(x)\,e^{i\omega t},\quad \theta=\Theta(\mathrm{x},\mathrm{t})+\mathrm{T}(\mathrm{x})\,\mathrm{e}^{\mathrm{i}\omega\mathrm{t}}, \tag{2.12}$$

where (U_r,Θ) satisfies the above initial boundary value problem with null boundary conditions and appropriate initial conditions, while (v_r,T) satisfies the following boundary value problem

$$(C_{rskl}v_{k,l}+M_{rs}T)_{,r}+\rho\omega^2 v_s=0,\quad \mathrm{x}\in\mathrm{B}, \tag{2.13}$$

$$(K_{rs}T_{,s})_{,r}+i\omega\theta_0 M_{rs}v_{r,s}=ci\omega T,\quad \mathrm{x}\in\mathrm{B}, \tag{2.14}$$

$$v_r(x)=0,\ \mathrm{T}(\mathrm{x})=0,\quad \mathrm{x}\in\partial\mathrm{D}\times[0,\mathrm{L}], \tag{2.15}$$

$$v_r(x)=f_r(x_1,x_2),\quad \mathrm{T}(\mathrm{x})=\mathrm{g}(\mathrm{x}_1,\mathrm{x}_2),\quad \mathrm{x}\in\mathrm{D}(0), \tag{2.16}$$

and

$$v_r(x)=0,\ \mathrm{T}(\mathrm{x})=0,\quad \mathrm{x}\in\mathrm{D}(\mathrm{L}), \tag{2.17}$$

in the case where L is finite.

We note that the spatial behavior of the transient solution (U_r,Θ) can be described by the methods developed in [15, 16]. The spatial behavior of the amplitude (v_r,T) of the forced oscillation has been established in [11] for the case of isotropic thermoelastic materials, provided the exciting frequency is less than a certain critical frequency.

In what follows we will establish some spatial estimates describing how the amplitude evolves with respect to the axial distance to the excited end under the assumption of strong ellipticity condition on the elasticity tensor for the low frequency range.

To this end we use the equations (2.13) and (2.14) and the lateral boundary condition (2.15) in order to obtain the following identity

$$2\int_{D(x_3)}\left(C_{rsmn}v_{r,s}\overline{v}_{m,n}-\rho\omega^2 v_s\overline{v}_s+\frac{c}{\theta_0}T\overline{T}+\delta K_{rs}T_{,r}\overline{T}_{,s}\right)da=$$

$$=\frac{d}{dx_3}\int_{D(x_3)}\left[(1-i\delta\omega\theta_0)\left(C_{3smn}v_{m,n}+M_{3s}T\right)\overline{v}_s+(1+i\delta\omega\theta_0)\left(C_{3smn}\overline{v}_{m,n}+M_{3s}\overline{T}\right)v_s+\right.$$

$$\left.+\left(\delta+\frac{i}{\omega\theta_0}\right)K_{3s}T\overline{T}_{,s}+\left(\delta-\frac{i}{\omega\theta_0}\right)K_{3s}\overline{T}T_{,s}\right]da, \tag{2.18}$$

where δ is a positive parameter at our disposal having the dimension of (time unit)/(temperatu unit). Moreover, a superposed bar denotes the complex conjugate of the corresponding quantity.

3 Spatial evolution in the low frequency range

We shall pursue here a method based on the identity (2.18). Such way proves that the critical frequency of the corresponding problem in classical elasticity is unaffected by the thermal effects. Thus, throughout this section we shall assume that the elasticity tensor is strongly elliptic and the conductivity tensor is positive. For convenience we will exemplify our method for the class of transversely isotropic elastic materials. Choosing the direction of transverse isotropy to coincide with the x_3 coordinate axis and using the standard notation

$$c_{ij}=C_{iijj},\quad \mathrm{i,j}\in\{1,2,3\},\quad (\mathrm{notsummed}),\quad \mathrm{c}_{44}=\mathrm{C}_{2323},\quad \mathrm{c}_{55}=\mathrm{C}_{1313},\quad \mathrm{c}_{66}=\mathrm{C}_{1212}, \tag{3.1}$$

the only non-zero coefficients of elasticity tensor are

$$c_{22}=\mathrm{c}_{11},\quad \mathrm{c}_{33},\quad \mathrm{c}_{23}=\mathrm{c}_{13},\quad \mathrm{c}_{44}=\mathrm{c}_{55},\quad \mathrm{c}_{66}=\frac{1}{2}\left(\mathrm{c}_{11}-\mathrm{c}_{12}\right). \tag{3.2}$$

Then the necessary and sufficient conditions for strong ellipticity to hold for a transversely isotropic material are (see, for example, [17-19])

$$c_{11}>0,\quad \mathrm{c}_{33}>0,\quad \mathrm{c}_{55}>0,\quad \mathrm{c}_{11}>\mathrm{c}_{12},\quad |\mathrm{c}_{13}+\mathrm{c}_{55}|<\mathrm{c}_{55}+\sqrt{\mathrm{c}_{11}\mathrm{c}_{33}}. \tag{3.3}$$

Moreover, the equations (2.13) and (2.14) become

$$\rho\omega^2 v_1+c_{11}v_{1,11}+(c_{12}+c_{66})\,v_{2,21}+c_{66}v_{1,22}+(c_{13}+c_{55})\,v_{3,31}+c_{55}v_{1,33}+M_{11}T_{,1}+M_{21}T_{,2}=0,$$

$$\rho\omega^2 v_2+(c_{12}+c_{66})\,v_{1,12}+c_{66}v_{2,11}+c_{11}v_{2,22}+(c_{13}+c_{55})\,v_{3,32}+c_{55}v_{2,33}+M_{12}T_{,1}+M_{22}T_{,2}=0,$$

$$\rho\omega^2 v_3+c_{55}\left(v_{3,11}+v_{3,22}\right)+(c_{13}+c_{55})\left(v_{1,1}+v_{2,2}\right)_{,3}+c_{33}v_{3,33}+M_{33}T_{,3}=0,$$

$$\left(K_{\alpha\beta}T_{,\beta}\right)_{,\alpha}+K_{33}T_{,33}+i\omega\theta_0\left(M_{\alpha\beta}v_{\alpha,\beta}+M_{33}v_{3,3}\right)=ci\omega T \tag{3.4}$$

and the identity (2.18) is

$$2\int_{D(x_3)}\left[c_{66}\left(v_{1,2}-v_{2,1}\right)\left(\overline{v}_{1,2}-\overline{v}_{2,1}\right)+c_{11}v_{\alpha,\alpha}\overline{v}_{\rho,\rho}+c_{33}v_{3,3}\overline{v}_{3,3}+\right.$$

$$+c_{13}\left(v_{\alpha,\alpha}\overline{v}_{3,3}+\overline{v}_{\alpha,\alpha}v_{3,3}\right)+c_{55}\left(v_{\alpha,3}\overline{v}_{3,\alpha}+\overline{v}_{\alpha,3}v_{3,\alpha}\right)+c_{55}\left(v_{\alpha,3}\overline{v}_{\alpha,3}+v_{3,\alpha}\overline{v}_{3,\alpha}\right)-$$

$$\left.-\varrho\omega^{2}v_{s}\overline{v}_{s}+\frac{c}{\theta_{0}}T\overline{T}+\delta K_{\alpha\beta}T_{,\alpha}\overline{T}_{,\beta}+\delta K_{33}T_{,3}\overline{T}_{,3}\right]da=$$

$$=\frac{d}{dx_{3}}\int_{D(x_{3})}\{(1-i\delta\omega\theta_{0})\left[c_{55}\overline{v}_{\alpha}\left(v_{3,\alpha}+v_{\alpha,3}\right)+\overline{v}_{3}\left(c_{13}v_{\alpha,\alpha}+c_{33}v_{3,3}\right)\right]+$$

$$+(1+i\delta\omega\theta_{0})\left[c_{55}v_{\alpha}\left(\overline{v}_{3,\alpha}+\overline{v}_{\alpha,3}\right)+v_{3}\left(c_{13}\overline{v}_{\alpha,\alpha}+c_{33}\overline{v}_{3,3}\right)\right]+$$

$$+K_{33}\left[\left(\delta+\frac{i}{\omega\theta_{0}}\right)T\overline{T}_{,3}+\left(\delta-\frac{i}{\omega\theta_{0}}\right)\overline{T}T_{,3}\right]\Bigg\}\,da. \tag{3.5}$$

On the other hand, by means of the boundary condition (2.15), we have

$$\frac{d}{dx_{3}}\int_{D(x_{3})}\left(v_{\alpha,\alpha}\overline{v}_{3}+\overline{v}_{\alpha,\alpha}v_{3}\right)da=\int_{D(x_{3})}\left(v_{\alpha,\alpha}\overline{v}_{3,3}+\overline{v}_{\alpha,\alpha}v_{3,3}\right)da-$$

$$-\int_{D(x_{3})}\left(v_{\alpha,3}\overline{v}_{3,\alpha}+\overline{v}_{\alpha,3}v_{3,\alpha}\right)da. \tag{3.6}$$

Now we introduce the function

$$I\left(x_{3}\right)=-\int_{D(x_{3})}\{(1-i\delta\omega\theta_{0})\left[c_{55}\overline{v}_{\alpha}\left(v_{3,\alpha}+v_{\alpha,3}\right)+\overline{v}_{3}\left(c_{13}v_{\alpha,\alpha}+c_{33}v_{3,3}\right)\right]+$$

$$+(1+i\delta\omega\theta_{0})\left[c_{55}v_{\alpha}\left(\overline{v}_{3,\alpha}+\overline{v}_{\alpha,3}\right)+v_{3}\left(c_{13}\overline{v}_{\alpha,\alpha}+c_{33}\overline{v}_{3,3}\right)\right]+$$

$$+2\varkappa\left(v_{\alpha,\alpha}\overline{v}_{3}+\overline{v}_{\alpha,\alpha}v_{3}\right)+K_{33}\left[\left(\delta+\frac{i}{\omega\theta_{0}}\right)T\overline{T}_{,3}+\left(\delta-\frac{i}{\omega\theta_{0}}\right)\overline{T}T_{,3}\right]\Bigg\}\,da \tag{3.6}$$

and note that the identities (3.4) and (3.5) imply

$$-\frac{dI}{dx_{3}}\left(x_{3}\right)=2\int_{D(x_{3})}\left[c_{66}\left(v_{1,2}-v_{2,1}\right)\left(\overline{v}_{1,2}-\overline{v}_{2,1}\right)-\varrho\omega^{2}v_{s}\overline{v}_{s}+\right.$$

$$+c_{11}v_{\alpha,\alpha}\overline{v}_{\rho,\rho}+\left(c_{13}+\varkappa\right)\left(v_{\alpha,\alpha}\overline{v}_{3,3}+\overline{v}_{\alpha,\alpha}v_{3,3}\right)+c_{33}v_{3,3}\overline{v}_{3,3}+$$

$$+c_{55}\left(v_{\alpha,3}\overline{v}_{\alpha,3}+v_{3,\alpha}\overline{v}_{3,\alpha}\right)-\left(c_{55}-\varkappa\right)\left(v_{\alpha,3}\overline{v}_{3,\alpha}+\overline{v}_{\alpha,3}v_{3,\alpha}\right)+$$

$$\left.+\frac{c}{\theta_{0}}T\overline{T}+\delta K_{\alpha\beta}T_{,\alpha}\overline{T}_{,\beta}+\delta K_{33}T_{,3}\overline{T}_{,3}\right]da. \tag{3.7}$$

In the above relations $\varkappa$ is a positive parameter at our disposal.

Furthermore, in view of the strong ellipticity condition (3.3), we can choose the parameter $\varkappa$ in such way to have

$$\max\left(-c_{13}-\sqrt{c_{11}c_{33}},0\right)<\varkappa<\min\left(2c_{55},-c_{13}+\sqrt{c_{11}c_{33}}\right), \tag{3.8}$$

so that we have

$$\left|c_{13}+\varkappa\right|<\sqrt{c_{11}c_{33}},\quad\left|c_{55}-\varkappa\right|<c_{55}. \tag{3.9}$$

On this basis we can conclude that

$$c_{11}v_{\alpha,\alpha}\overline{v}_{\rho,\rho}+\left(c_{13}+\varkappa\right)\left(v_{\alpha,\alpha}\overline{v}_{3,3}+\overline{v}_{\alpha,\alpha}v_{3,3}\right)+c_{33}v_{3,3}\overline{v}_{3,3}\geq$$

$$\geq \chi_1 \left(v_{\alpha,\alpha} \overline{v}_{\rho,\rho} + v_{3,3} \overline{v}_{3,3} \right), \tag{3.10}$$

$$c_{55} \left(v_{\alpha,3} \overline{v}_{\alpha,3} + v_{3,\alpha} \overline{v}_{3,\alpha} \right) + \left(c_{55} - \varkappa \right) \left(v_{\alpha,3} \overline{v}_{3,\alpha} + \overline{v}_{\alpha,3} v_{3,\alpha} \right) \geq$$

$$\geq \chi_2 \left(v_{\alpha,3} \overline{v}_{\alpha,3} + v_{3,\alpha} \overline{v}_{3,\alpha} \right), \tag{3.11}$$

where

$$\chi_1 = \frac{1}{2} \left(c_{11} + c_{33} - \sqrt{\left(c_{11} - c_{33} \right)^2 + 4 \left(c_{13} + \varkappa \right)^2} \right), \tag{3.12}$$

$$\chi_2 = \min \left(\varkappa, 2c_{55} - \varkappa \right). \tag{3.13}$$

On the other hand, in view of the boundary condition (2.15), we have

$$\int_{D(x_3)} v_{\alpha,\beta} \overline{v}_{\alpha,\beta} da \geq \lambda \int_{D(x_3)} v_\alpha \overline{v}_\alpha da, \tag{3.14}$$

$$\int_{D(x_3)} v_{3,\alpha} \overline{v}_{3,\alpha} da \geq \lambda \int_{D(x_3)} v_3 \overline{v}_3 da, \tag{3.15}$$

where $\lambda > 0$ is the first eigenvalue in the two–dimensional clamped membrane eigenvalue problem for the cross section $D(x_3)$.

Throughout the remainder of this section we will assume that

$$0 < \omega < \omega_1, \tag{3.16}$$

where

$$\omega_1 = \sqrt{\frac{\lambda}{\varrho} \min \left(\min \left(c_{66}, \chi_1 \right), \chi_2 \right)}. \tag{3.17}$$

Consequently, the relations (3.7) to (3.16) give

$$-\frac{dI}{dx_3}(x_3) \geq 2 \min \left(c_{66}, \chi_1 \right) \left(1 - \frac{\omega^2}{\omega_1^2} \right) \int_{D(x_3)} v_{\alpha,\beta} \overline{v}_{\alpha,\beta} da + \chi_1 \int_{D(x_3)} v_{3,3} \overline{v}_{3,3} da +$$

$$+ \chi_2 \int_{D(x_3)} v_{\alpha,3} \overline{v}_{\alpha,3} da + \chi_2 \left(1 - \frac{\omega^2}{\omega_1^2} \right) \int_{D(x_3)} v_{3,\alpha} \overline{v}_{3,\alpha} da +$$

$$+ \int_{D(x_3)} \left(\frac{c}{\theta_0} T \overline{T} + \delta K_{\alpha\beta} T_{,\alpha} \overline{T}_{,\beta} + \delta K_{33} T_{,3} \overline{T}_{,3} \right) da. \tag{3.18}$$

In what follows we will discuss the following hypotheses upon c and K_{rs}: (A) $c > 0$, $K_{33} > 0$ and $K_{\alpha\beta} \xi_\alpha \xi_\beta \geq 0$ for all real vectors ξ_α and (B) $c \leq 0$, $K_{33} > 0$ and $K_{\alpha\beta} \xi_\alpha \xi_\beta \geq k_0 \xi_\alpha \xi_\alpha$ for all real vectors ξ_α.

It is easy to see that in the case of hypothesis (A) we have

$$-\frac{dI}{dx_3}(x_3) \geq 2 \min \left(c_{66}, \chi_1 \right) \left(1 - \frac{\omega^2}{\omega_1^2} \right) \int_{D(x_3)} v_{\alpha,\beta} \overline{v}_{\alpha,\beta} da + \chi_1 \int_{D(x_3)} v_{3,3} \overline{v}_{3,3} da +$$

$$+ \chi_2 \int_{D(x_3)} v_{\alpha,3} \overline{v}_{\alpha,3} da + \chi_2 \left(1 - \frac{\omega^2}{\omega_1^2} \right) \int_{D(x_3)} v_{3,\alpha} \overline{v}_{3,\alpha} da +$$

$$+\frac{c}{\theta_0}\int_{D(x_3)} T\overline{T}da + \delta K_{33}\int_{D(x_3)} T_{,3}\overline{T}_{,3}da \geq 0, \tag{3.19}$$

for all $\delta > 0$. For the hypothesis (B) we use the inequality

$$\int_{D(x_3)} T_{,\alpha}\overline{T}_{,\alpha}da \geq \lambda\int_{D(x_3)} T\overline{T}da, \tag{3.20}$$

and choose the parameter δ in such way to have

$$\delta > -\frac{c}{k_0\theta_0\lambda}. \tag{3.21}$$

We further set

$$\delta_0 = \delta k_0\lambda + \frac{c}{\theta_0} \tag{3.22}$$

and note that (3.19) gives

$$-\frac{dI}{dx_3}(x_3) \geq 2\min(c_{66},\chi_1)\left(1-\frac{\omega^2}{\omega_1^2}\right)\int_{D(x_3)} v_{\alpha,\beta}\overline{v}_{\alpha,\beta}da + \chi_1\int_{D(x_3)} v_{3,3}\overline{v}_{3,3}da+$$

$$+\chi_2\int_{D(x_3)} v_{\alpha,3}\overline{v}_{\alpha,3}da + \chi_2\left(1-\frac{\omega^2}{\omega_1^2}\right)\int_{D(x_3)} v_{3,\alpha}\overline{v}_{3,\alpha}da+$$

$$+\delta_0\int_{D(x_3)} T\overline{T}da + \delta K_{33}\int_{D(x_3)} T_{,3}\overline{T}_{,3}da \geq 0. \tag{3.23}$$

Concluding, for the both hypotheses in study we can deduce that $I(x_3)$ is a non-increasing function with respect to x_3 on $[0, L]$. Therefore, on the basis of the end condition (2.17), from the relation (3.6) we get $I(L) = 0$ and hence

$$I(x_3) \geq 0 \quad \text{forall } \mathrm{x}_3 \in [0, \mathrm{L}]. \tag{3.24}$$

By integrating the relations (3.19) and (3.23) with respect to the x_3 variable on $[x_3, L]$ we get

$$I(x_3) \geq 2\min(c_{66},\chi_1)\left(1-\frac{\omega^2}{\omega_1^2}\right)\int_{Bx_3} v_{\alpha,\beta}\overline{v}_{\alpha,\beta}dv + \chi_1\int_{Bx_3} v_{3,3}\overline{v}_{3,3}dv+$$

$$+\chi_2\int_{Bx_3} v_{\alpha,3}\overline{v}_{\alpha,3}dv + \chi_2\left(1-\frac{\omega^2}{\omega_1^2}\right)\int_{Bx_3} v_{3,\alpha}\overline{v}_{3,\alpha}dv+$$

$$+\frac{c}{\theta_0}\int_{Bx_3} T\overline{T}dv + \delta K_{33}\int_{Bx_3} T_{,3}\overline{T}_{,3}dv \geq 0, \tag{3.25}$$

for the hypothesis (A) and

$$I(x_3) \geq 2\min(c_{66},\chi_1)\left(1-\frac{\omega^2}{\omega_1^2}\right)\int_{Bx_3} v_{\alpha,\beta}\overline{v}_{\alpha,\beta}dv + \chi_1\int_{Bx_3} v_{3,3}\overline{v}_{3,3}dv+$$

$$+\chi_2\int_{Bx_3} v_{\alpha,3}\overline{v}_{\alpha,3}dv + \chi_2\left(1-\frac{\omega^2}{\omega_1^2}\right)\int_{Bx_3} v_{3,\alpha}\overline{v}_{3,\alpha}dv+$$

$$+\delta_0 \int_{Bx_3} T\overline{T}dv + \delta K_{33} \int_{Bx_3} T_{,3}\overline{T}_{,3}dv \geq 0. \tag{3.26}$$

As we can see, the relations (3.25) and (3.26) prove that $I(x_3)$ represents a measure of the amplitude of vibration.

Further, we can use the Schwarz and arithmetic–geometric mean inequalities and the relations (3.25) and (3.26) to estimate $I(x_3)$ in terms of the derivative $\frac{dI}{dx_3}(x_3)$. So we are led to the following first order differential inequality

$$\frac{dI}{dx_3}(x_3) + \sigma I(x_3) \leq 0 \quad \text{forall } x_3 \in [0, L], \tag{3.27}$$

which integrated furnishes the exponential spatial decay

$$0 \leq I(x_3) \leq I(0)\, e^{-\sigma x_3} \quad \text{forall } x_3 \in [0, L]. \tag{3.28}$$

Here σ is a computable positive constant depending on $\varkappa$, δ, ω, ω_1 and the termoelastic coefficients.

4 A semi-infinite cylinder

The foregoing analysis can be extended to the case of a semi-infinite cylinder (that is the case when $L \to \infty$) getting some alternatives of Phragmén–Lindelöf type. In this case we associate with the amplitude (v_r, T) the following energetic measure

$$E(x_3) = 2\int_{B_{x_3}} \left[c_{66}(v_{1,2} - v_{2,1})(\overline{v}_{1,2} - \overline{v}_{2,1}) - \varrho\omega^2 v_s \overline{v}_s + \right.$$
$$+ c_{11} v_{\alpha,\alpha}\overline{v}_{\rho,\rho} + (c_{13} + \varkappa)(v_{\alpha,\alpha}\overline{v}_{3,3} + \overline{v}_{\alpha,\alpha}v_{3,3}) + c_{33}v_{3,3}\overline{v}_{3,3} +$$
$$+ c_{55}(v_{\alpha,3}\overline{v}_{\alpha,3} + v_{3,\alpha}\overline{v}_{3,\alpha}) + (c_{55} - \varkappa)(v_{\alpha,3}\overline{v}_{3,\alpha} + \overline{v}_{\alpha,3}v_{3,\alpha}) +$$
$$\left. + \frac{c}{\theta_0} T\overline{T} + \delta K_{\alpha\beta} T_{,\alpha}\overline{T}_{,\beta} + \delta K_{33} T_{,3}\overline{T}_{,3}\right] dv, \tag{4.1}$$

where the positive parameters $\varkappa$ and δ satisfy relations (3.8) and (3.21) (when $c < 0$), respectively. In view of the fact that $I(x_3)$ is a non-increasing function on $[0, \infty)$ it results to be possible only the two possibilities: (i) $I(x_3) \geq 0$ for all $x_3 \in [0, \infty)$; or (ii) there exists $x_3^* \in [0, \infty)$ so that $I(x_3^*) < 0$.

In the case (i) we can proceed in a manner similar with that for obtaining the decay estimate (3.28). If we use such a result and integrate the inequality (3.7) on $[0, \infty)$, then we obtain

$$I(x_3) = E(x_3) \tag{4.2}$$

and hence the energetic measure $E(x_3)$ of the amplitude of harmonic vibration exists and it satisfies the following spatial decay inequality

$$0 \leq E(x_3) \leq E(0)\, e^{-\sigma x_3} \quad \text{forall } x_3 \in [0, \infty). \tag{4.3}$$

Let us discuss the case (ii). Then it follows that

$$I(x_3) < 0 \quad \text{forall } x_3 \in [x_3^*, \infty). \tag{4.4}$$

Furthermore, the relation (3.23) and the Schwarz and arithmetic–geometric mean inequalities give now

$$\frac{dI}{dx_3}(x_3) - \sigma I(x_3) \leq 0 \quad \text{forall } x_3 \in [0, \infty), \tag{4.5}$$

which implies

$$-I(x_3) \geq I(x_3^*)\, e^{\sigma(x_3 - x_3^*)} \quad \text{forall } x_3 \in [x_3^*, \infty). \tag{4.6}$$

Thus, we have $I(\infty) \equiv \lim_{x_3 \to \infty} I(x_3) = -\infty$ and from relations (3.7) and (4.1) we deduce that the energetic measure $E(x_3)$ is infinite.

Concluding, we have obtained the following alternative of Phragmén-Lindelöf type result: within the context of a semi-infinite cylinder made of a transversely isotropic material the amplitude (v_r, T) either has a finite energetic measure $E(x_3)$ that satisfies the spatial decay estimate (4.3) or it has an infinite energetic measure and then $-I(x_3)$ goes to infinity faster than the exponential $e^{\sigma(x_3 - x_3^*)}$.

5 Some comments

The main purpose of this paper was to develop a method for studying the spatial behavior of the harmonic vibrations in linear thermoelasticity for the class of transversely isotropic materials, but it can be equally pursued for the class of rhombic materials in view of the characterization of the strong ellipticity established in [19].

The foregoing analysis is valid, however, in the low frequency range. It should be mentioned that the present estimates fall to give information regarding the spatial behavior for harmonic vibrations with frequencies close to the critical value. Although considerable progress has been made on the subject, it is apparent that an open question remain to be fully resolved for high frequencies. In the high frequency range it should be expected to have spatial decay slower than that for low frequency range. Such analysis will be pursued in a future research study.

Acknowledgement

The work by S.C. was supported by Romanian Ministry of Education and Research, CNCSIS Grant CERES 2-CEex06-11-12/25.07.2006.

References

[1] A.-J.-C. B. de Saint-Venant, Mémoire sur la torsion des prismes. *Mémoires présentés pars divers Savants a l'Académie des Sciences de l'Institut Impérial de France,* vol. **14** (1853), pp. 233–560.

[2] A.-J.-C. B. de Saint-Venant, Mémoire sur la flexion des prismes. *J. Math. Pures Appl.,* vol. **1** (Ser. 2) (1856), pp. 89–189.

[3] B.A. Boley, Application of Saint–Venant's principle in dynamical problems. *J. Appl. Mech.,* vol. **22** (1955), pp. 204–206.

[4] B.A. Boley, Some observations on Saint–Venant's principle. In: *Proc. 3rd U.S. Nat. Cong. Appl. Mech.*, ASME, New York, 1958, pp. 259–264.

[5] C.O. Horgan and J.K. Knowles, Recent developments concerning Saint–Venant's principle. In: *Advances in Applied Mechanics (T.Y. Wu and J.W. Hutchinson, eds.)*, vol. 23, Academic Press, New York, 1983, pp. 179–269.

[6] C.O. Horgan, Recent developments concerning Saint–Venant's principle: An update. Appl. Mech. Rev., vol. **42** (1989), pp. 295–303.

[7] C.O. Horgan, Recent developments concerning Saint–Venant's principle: A second update. Appl. Mech. Rev., vol. **49** (1996), pp. S101–S111.

[8] J.N. Flavin and R.J. Knops, Spatial decay estimates in continuum dynamics. *J. Elasticity,* vol. **17** (1987), pp. 249–264.

[9] J.N. Flavin, R.J. Knops and L.E. Payne, Energy bounds in dynamical problems for a semi-infinite elastic beam. In: *Elasticity, mathematical methods and applications (G. Eason and R.W. Ogden, eds.),* Ellis–Horwood, Chichester, 1990, pp. 101–111.

[10] R.J. Knops, Spatial decay estimates in the vibrating anisotropic elastic beam. In: *Waves and Stability in Continuous Media (S. Rionero, ed.),* World Scientific, Singapore, 1991, pp. 192–203.

[11] S. Chiriţă, Spatial decay estimates for solutions describing harmonic vibrations in a thermoelastic cylinder. *J. Thermal Stresses,* vol. **18** (1995), pp. 421–436.

[12] S. Chiriţă, Some spatial decay estimates for the harmonic vibrations in anisotropic linear thermoelasticity. *Bull. Techn. Univ. Istanbul,* vol. **48** (1996), pp. 537-549.

[13] M. Aron and S. Chiriţă, Decay and continuous dependence estimates for harmonic vibrations of micropolar elastic cylinders. *Arch. Mechanics,* vol. **49** (1997), pp. 665–675.

[14] D. R. Carlson, Linear thermoelasticity. In: Truesdell, C. (ed.) Handbuch der Physik, vol. VIa/2. Springer, Berlin Heidelberg New York, (1972).

[15] S. Chiriţă and M. Ciarletta, Time-weighted surface power function method for the study of spatial behaviour in dynamics of continua. *Eur. J. Mech. A/Solids,* vol. **18** (1999), pp. 915–933.

[16] S. Chiriţă, Spatial behavior in the strongly elliptic anisotropic thermoelastic materials. *J. Thermal Stresses,* vol. **30** (2007), pp. 859–873.

[17] J. Merodio and R.W. Ogden, A note on strong ellipticity for transversely isotropic linearly elastic solids. *Quart. J. Mech. Appl. Math.,* vol. **56** (2003), pp. 589-591.

[18] S. Chiriţă, On the strong ellipticity condition for transversely isotropic linearly elastic solids. *An. St. Univ. Iasi, Matematica,* vol. **52**, f. 2, (2006), pp. 245-250.

[19] S. Chiriţă, A. Danescu and M. Ciarletta, On the strong ellipticity of the anisotropic linearly elastic materials. *J. Elasticity,* vol. **87**, (2007), pp. 109–132.

On the Nonautonomous Lotka-Volterra System

R. De Luca, S. Rionero

Department of Mathematics and Applications *R. Caccioppoli*,
University of Naples *Federico II*,
Complesso Universitario Monte S.Angelo, via Cinzia, 80126 Naples, Italy.

1 Introduction

The perturbed Lotka-Volterra bidimensional predator-prey model has been deeply developed by many authors (see, for instance [1]-[14] and the reference therein). Different types of perturbations terms have been introduced in order to account for (small) variations of the idealized hypotheses of Lotka-Volterra models as well as to put controls on the growth of both predator and prey. In [8], [9], [10], small autonomous perturbations of very general type, depending on a small parameter, have been considered and their influence on the stability of the positive equilibrium point of the classical Lotka-Volterra model, or on the existence both of periodic solutions of the perturbed system or perturbed critical points. In [4] time dependent perturbation terms, depending on the small parameter, have been introduced in the classic model and their influence on the stability of the equilibria of this model has been studied by using the multiple scale method. In [5], [6], Fergola Rionero Tenneriello, consider the same type of nonautonomous perturbation, but they do not make any assumption on the *smallness* of these perturbations. Precisely, they consider the following system

$$\begin{cases} \dot{x} = ax - bxy + D(t)(y - x) \\ \\ \dot{y} = -cy + dxy + D(t)(x - y) \end{cases} \tag{1}$$

with x and y respectively prey and predator densities and a, b, c, d positive constants such that $\frac{c}{d} = \frac{a}{b}$ and $D = D(t)$ regular enough to assure global (in time) existence and uniqueness of the solutions. Using the Liapunov Direct Method they obtain conditions assuring exponential stability of

$$\overline{x} = \frac{c}{d} \ , \ \overline{y} = \frac{a}{b}$$

under the fundamental assumption that $\inf_{t \in \mathcal{R}^+} D(t) = D_0 > 0$.

In [14] this restriction has been eliminated and substituted by larger conditions on

$$\int_0^t D(\tau) d\tau \tag{2}$$

In this paper -following [15]- we consider the two perturbed Lotka-Volterra systems

$$\begin{cases} \dot{x} = f(t)(a - by)x \\ \\ \dot{y} = g(t)(-c + dx)y \end{cases} \tag{3}$$

$$\begin{cases} \dot{x} = (a - by)\,x + D(t)\left(\frac{cb}{ad}y - x\right) \\ \\ \dot{y} = (-c + dx)\,y + D(t)\left(\frac{ad}{bc}x - y\right) \end{cases} \tag{4}$$

with a, b, c, d positive constants and

$$f,\, g,\, D : \mathcal{R}^{+} \to \mathcal{R} \tag{5}$$

sufficiently smooth for guaranteeing global existence and uniqueness of the solutions.

Remark 1. *We remark that:*

(i) Both the models (3)-(4) admit the critical point

$$\overline{x} = \frac{c}{d} \quad , \quad \overline{y} = \frac{a}{b}$$

of the classical Lotka-Volterra system.

(ii) The temporal functions f, g, D can be interpreted as perturbations terms or controls.

2 Stability of $(\overline{x}, \overline{y})$ through model (3)

Setting

$$\begin{cases} x = u + \overline{x} \\ \\ y = v + \overline{y} \end{cases} \tag{6}$$

it turns out that

$$\begin{cases} \dot{u} = -\overline{x}bfv - bfuv \\ \\ \dot{v} = \overline{y}dgu + dguv \end{cases} \tag{7}$$

By using the Liapunov function introduced by Rionero [15-16]

$$V = \frac{1}{2}\left[A\left(u^2 + v^2\right) + d^2\overline{y}^2 g^2 u^2 + b^2 \overline{x}^2 f^2 v^2\right] \tag{8}$$

with

$$A = \begin{vmatrix} 0 & -\overline{x}bf \\ \overline{y}dg & 0 \end{vmatrix} \tag{9}$$

it turns out that the time derivative of V along the solutions of (7) is

$$\dot{V} = \frac{1}{2}\left[ac\frac{d}{dt}(fg)\left(u^2+v^2\right) + 2d^2\overline{y}^2 g\dot{g}u^2 + 2b^2\overline{x}^2 f\dot{f}v^2\right] + \psi \tag{10}$$

where

$$\begin{aligned}\psi &= -\left(A + d^2\overline{y}^2 g^2\right) bfu^2v + \left(A + b^2\overline{x}^2 f^2\right) dguv^2 = \\ &= -\left(acf + d^2\overline{y}^2 g\right) bfgu^2v + \left(acg + b^2\overline{x}^2 f\right) dfguv^2\end{aligned} \tag{11}$$

In view of (10), one obtains

$$\dot{V} = \frac{1}{2}\left[u^2\frac{d}{dt}\left(acfg + d^2\overline{y}^2 g^2\right) + v^2\frac{d}{dt}\left(acfg + b^2\overline{x}^2 f^2\right)\right] + \psi \tag{12}$$

Setting

$$F = \frac{1}{2}\left(acfg + d^2\overline{y}^2 g^2\right) \qquad G = \frac{1}{2}\left(acfg + b^2\overline{x}^2 f^2\right) \tag{13}$$

it follows that

$$\begin{cases} V = Fu^2 + Gv^2 \\ \dot{V} = \dot{F}u^2 + \dot{G}v^2 + \psi \end{cases} \tag{14}$$

Theorem 1. *Let*

$$\begin{cases} A > h_1 \qquad \forall t \geq 0 \\ \dot{F} \leq -h_2 \\ \dot{G} \leq -h_2 \end{cases} \tag{15}$$

with f, g bounded and h_i (i=1,2) positive constants. Then $(\overline{x}, \overline{y})$ is locally asymptotically exponentially stable.

Proof. Obviously, since $A > h_1 > 0$, V is positive definite, while

$$\dot{V} = \dot{F}u^2 + \dot{G}v^2 + \psi \leq -h_2\left(u^2+v^2\right) + \psi \tag{16}$$

Further

$$\psi = -2Ffbu^2v + 2Ggduv^2 \tag{17}$$

and exists a constant m such that

$$|Ffb| + |Ggd| \leq m \tag{18}$$

Since $u^2|v| + |u|v^2 \leq \frac{\sqrt{2}}{2}\left(u^2+v^2\right)^{\frac{3}{2}}$, setting

$$E = u^2 + v^2 \tag{19}$$

one obtains

$$\dot{V} \leq -\left(h_2 - m_1 E^{\frac{1}{2}}\right) E \tag{20}$$

with $m_1 = m\sqrt{2}$. Therefore

$$E_0^{\frac{1}{2}} = \delta < \frac{h_2}{m_1} \tag{21}$$

implies (by recursive arguments)

$$\dot{V} \leq -\left(h_2 - m_1 E_0^{\frac{1}{2}}\right) E \leq -(h_2 - \delta)E$$

Since V is equivalent to E, exist two positive constants K_i (i=1,2), such that

$$K_1 E \leq V \leq K_2 E \tag{22}$$

and hence

$$\dot{V} \leq -\frac{(h_2 - \delta)}{K_2} V \tag{23}$$

i.e.

$$V \leq V_0 exp\left[-\frac{h_2 - \delta}{K_2} t\right] \tag{24}$$

Remark 2. *An immediate interesting application which can be interpreted as a control on initial perturbation of the model, can be obtained by choosing*

$$\begin{cases} f = 1 \pm \varepsilon_1 exp^{-K_1 t} \\ g = 1 \pm \varepsilon_2 exp^{-K_2 t} \end{cases}$$

or

$$\begin{cases} f = 1 \pm \varepsilon_1 exp^{-K_1 t} \\ g = 1 \mp \varepsilon_2 exp^{-K_2 t} \end{cases}$$

with $0 < \varepsilon_i < 1$, $K_i > 0$ *(i=1,2).*

3 Stability of $(\overline{x}, \overline{y})$ through model (4)

Model (4) is:

$$\begin{cases} \dot{x} = (a - by)\, x + D(t) \left(\frac{cb}{ad} y - x\right) \\ \dot{y} = (-c + dx)\, y + D(t) \left(\frac{ad}{bc} x - y\right) \end{cases}$$

Setting

$$\begin{cases} x = u + \overline{x} \\ y = v + \overline{y} \end{cases}$$

it turns out that

$$\begin{cases} \dot{u} = -D(t)u + \frac{bc}{ad}\left(D(t) - a\right)v - buv \\ \\ \dot{v} = \frac{ad}{bc}\left(D(t) + c\right)u - D(t)v + duv \end{cases} \tag{25}$$

By using the Liapunov functional, {Rionero [15-16]}

$$V = \frac{1}{2}\left[A\left(u^2 + v^2\right) + \left(D(t)v + \frac{ad}{bc}\left(D(t) + c\right)u\right)^2 + \left(D(t)u + \frac{bc}{ad}\left(D(t) - a\right)v\right)^2\right] \tag{26}$$

with

$$A = \begin{vmatrix} -D(t) & \frac{bc}{ad}\left(D(t) - a\right) \\ \\ \frac{ad}{bc}\left(D(t) + c\right) & -D(t) \end{vmatrix} \tag{27}$$

and

$$I(t) = -2D(t) \tag{28}$$

it turns out that

$$\begin{aligned} \dot{V} &= \frac{1}{2}\left[\left(AI + \frac{d}{dt}\left(A + \frac{a^2d^2}{b^2c^2}\left(D + c\right)^2 + D^2\right)\right)u^2\right] + \\ &+ \frac{1}{2}\left[\left(AI + \frac{d}{dt}\left(A + \frac{b^2c^2}{a^2d^2}\left(D - a\right)^2 + D^2\right)\right)v^2\right] + \psi \end{aligned} \tag{29}$$

with

$$\begin{cases} \psi &= \left(\alpha_1 u - \alpha_3 v\right)\left(-buv\right) + \left(\alpha_2 v - \alpha_3 u\right)\left(duv\right) \\ \\ \alpha_1 &= A + \frac{a^2d^2}{b^2c^2}\left(D + c\right)^2 + D^2 \\ \\ \alpha_2 &= A + \frac{b^2c^2}{a^2d^2}\left(D - a\right)^2 + D^2 \\ \\ \alpha_3 &= -D\left[\frac{ad}{bc}(D + c) + \frac{bc}{ad}(D - a)\right] \end{cases} \tag{30}$$

Theorem 2. *Let D be bounded and either*

$$\begin{cases} a > c, \ D(t) > \dfrac{ac}{c-a} \\ \\ -D\left(D\left(a - c\right) + ac\right) + \frac{1}{2}\frac{d}{dt}\left(D\left(a - c\right) + ac + \frac{a^2d^2}{b^2c^2}\left(D + c\right)^2 + D^2\right) \leq -h \\ \\ -D\left(D\left(a - c\right) + ac\right) + \frac{1}{2}\frac{d}{dt}\left(D\left(a - c\right) + ac + \frac{b^2c^2}{a^2d^2}\left(D - a\right)^2 + D^2\right) \leq -h \end{cases} \tag{31}$$

or

$$\begin{cases} a < c, \ D(t) < \frac{ac}{c-a} \\ -D\left(D\left(a-c\right)+ac\right)+\frac{1}{2}\frac{d}{dt}\left(D\left(a-c\right)+ac+\frac{a^2d^2}{b^2c^2}\left(D+c\right)^2+D^2\right) \quad \leq \quad -h \\ -D\left(D\left(a-c\right)+ac\right)+\frac{1}{2}\frac{d}{dt}\left(D\left(a-c\right)+ac+\frac{b^2c^2}{a^2d^2}\left(D-a\right)^2+D^2\right) \quad \leq \quad -h \end{cases} \tag{32}$$

or

$$\begin{cases} a = c \\ -Da^2+\frac{1}{2}\frac{d}{dt}\left(\frac{d^2}{b^2}\left(D+a\right)^2+D^2\right) \quad \leq \quad -h \\ -Da^2+\frac{1}{2}\frac{d}{dt}\left(\frac{b^2}{d^2}\left(D-a\right)^2+D^2\right) \quad \leq \quad -h \end{cases} \tag{33}$$

where h is a positive constant. Then $(\overline{x},\overline{y})$ is locally asymptotically exponentially stable.

Proof. Conditions 31.1, 32.1 assure that $A > 0$, while in the case $a = c$, $A = a^2 > 0$. So V is positive definite. The other two conditions guarantee that

$$\dot{V} \leq -h\left(u^2+v^2\right)+\psi$$

Further

$$\psi = -\left(b\alpha_1+\alpha_3 d\right)u^2 v+\left(b\alpha_3+\alpha_2 d\right)uv^2$$

Since $D(t)$ is bounded, a constant m exists such that

$$|b\alpha_1+\alpha_3 d|+|b\alpha_3+\alpha_2 d| \leq m$$

Following the same argument used in proof of Theorem 1, the thesis is hold.

Remark 3. *An immediate application in the case $a = c$ can be obtained by choosing*

$$D(t) = h^* + f(t)$$

with $h^* > max\left\{\frac{h}{a^2}, \frac{2b^2a}{(1+b^2)}\right\}$ and

$$\begin{array}{ll} f(t) > 0 & \forall t > 0 \\ \dot{f}(t) < 0 & \forall t > 0 \end{array}$$

References

[1] Comins H.N., Blatt D.W.E., *Prey predator models in spatially heterogeneous environments.* J. Theor. Biol (1974), 75-83.

[2] Dutt R., Glosh P.K., *Nonlinear correction to Lotka Volterra oscillations in a prey predator system.* Math. Biosci **27** (1975) 9-16.

[3] Dutt R., Glosh P.K., Karmaker B.B., *Application of perturbation theory to the nonlinear Volterra Gause Witt model for prey predator interaction.* Bull. Math. Biol **37** (1975), 139-146.

[4] Fergola P., Nayfeh A.H., *Application of the method of multiple scales to perturbed Lotka Volterra models.* Procedings of Vth International Conference "Waves and Stability in continuous media" S. Giambo', Editel Cosenza (1987).

[5] Fergola P., Rionero S., Tenneriello C., *Asymptotic stability of a perturbed Lotka Volterra system.* Procedings of Vth International Conference "Waves and Stability in continuous media" Series and Advances in Mathematics for Applied Sciences **4** (1990).

[6] Fergola P., Rionero S., Tenneriello C., *On the stability of Lotka Volterra perturbed equations.* Atti Acc. Peloritana Pericolanti Cl. Sci. Fis. Natur. **48** (1991), 235-256.

[7] Freedman H.I., *Deterministic mathematical models in population ecology.* Marcel Dekker, Inc. New York (1980).

[8] Freedman H.I., Waltman P., *Perturbation of two dimensional predator prey equations.* Siam J. Appl. Math. **28** (1975), 1-10.

[9] Freedman H.I., Waltman P., *Periodic solutions of perturbed Lotka Volterra systems.* International Conference on Differential Equations, H.A. Antosiewicz Academic Press New York (1975), 312-316.

[10] Freedman H.I., Waltman P., *Perturbation of two dimensional predator prey equations with an unperturbed critical point.* **29** Siam J. Appl. Math (1975), 719-733.

[11] Levin S.A., *Dispersion and population interactions.* Amer. Nat **108** (1974), 207-228.

[12] Samuelson P.A., *Generalized predator prey oscillations in ecological and economic equilibrium.* Proc. Nat. Acad. Sei. USA **68** (1971), 980-983.

[13] Walter C., *The global asymptotic stability of prey predator systems with second order dissipation.* Bull. Math. Biol **36** (1974), 215-217.

[14] Capone F., Rionero S., *Attractivity conditions for a perturbed Lotka Volterra model.* Procedings of VIth International Conference "Waves and Stability in continuous media" **46** (1991), 55-63.

[15] Rionero S., *On the stability of a nonautonomous binary system of O.D.E.* (submitted).

[16] Rionero S., *L^2 stability of the solutions to a nonlinear binary reaction-diffusion system of P.D.Es.* Rendiconti Lincei, s.9 ,**16**, 227-238 (2005)

[17] Rionero S., *A rigorous reduction of the stability of the solutions to a nonlinear binary reaction-diffusion system of P.D.Es. to the stability of the solutions to a linear binary system of O.D.Es.* J.Math Anal. Appl., **319**, 377-397

Structure Order Balance Law and Phase Transitions

*M. Fabrizio**

Department of Mathematics,
Bologna, Italy.

Abstract

In this paper, by means of a new balance law on the structure order, the general Ginzburg-Landau model used in superconductivity is extend to first order phase transitions. In particular we shall apply this model to the study of the liquid-vapor transition.

1 Introduction

The Ginzburg-Landau [7] equation and the following dynamical extension presented by Gor'kov and Eliashberg [8], are able to study the phase transition which occur in a metal or alloy superconductor. In [3] we have observed that this equation can be obtained from a balance law on the structure order and may be used for a description of any phase transition. In this work we have studied the transition of first order which we meet for liquid-vapor phase transition. For this phenomenon we present a mathematical model capable of describing a phase transition by means of an *order parameter* f, such that $0 \le f \le 1$, and by a decomposition of the *density* ρ into two components ρ_0 and ρ_1. The density ρ_0 is connected with the *pressure* p and hence any variation of ρ_0 involves a variation of the pressure. On the contrary, the variation of the component ρ_1, independent of the pressure, is due to the phase transition. In other words, ρ_1 becomes a function of the order parameter f.

Moreover, by the Second Law of Thermodynamics, we obtain the restrictions which bring to the relation between the density ρ_1 and the phase f. Finally, in the study of the liquid-vapor transition, the model is able to well represent the Andrews diagram which describes such a transition.

2 Balance law on the structure order

In this section we follow an approach to the description of the phase transition by means of a balance equation on the structure order. We begin to consider that in any phase transition the states of the two phases are endowed with different structures

*Email: `fabrizio@dm.unibo.it`

or symmetries. In particular for many materials the structure order below a critical temperature is greater than above. That is why in this paper we propose a view of the phase equation as a balance law on the order of the structure.

In this framework, consider a body $\mathcal{B}$ and let the mass density ρ be constant. For any sub-body $S \in \mathcal{B}$, we represent the balance of the structure order in the form

$$\int_S \rho k dv = \int_{\partial S} p \cdot n d\sigma + \int \rho\sigma dv$$

where k is the internal structure order per unit mass, p is the order vector flux, n is the outward normal to $\partial\Omega$ and σ is the order supply. The divergence theorem, the arbitrariness of S and the smoothness of $\rho k, \nabla \cdot p$ and $\rho\sigma$ allow us to write

$$\rho k = \nabla \cdot p + \rho\sigma \tag{1}$$

From this equation, by means of the order parameter f, we obtain the following balance equation on the power of the structure order

$$\rho k f_t = \nabla \cdot (p f_t) - p \cdot \nabla f_t + \rho\sigma f_t \tag{2}$$

For any sub-body S, from (2) we have

$$\int_S (\rho k f_t + p \cdot \nabla f_t) dv = \int_{\partial S} p f_t \cdot n d\sigma + \int_S \rho\sigma f_t dv$$

where the left-hand side denotes the interval power of the structure order, the first integral in the right-hand side denotes the power flux and the last integral is the external power supply. Therefore, in the following, we denote by

$$\mathcal{P}^i = \rho k f_t + p \cdot \nabla f_t, \quad \mathcal{P}^e = \nabla \cdot (p f_t) + \rho\sigma f_t \tag{3}$$

the internal and the external power density respectively. In order to study the phase transitions we suppose that the equation (1) assumes the Ginzburg-Landau representation. Hence,

$$k = \gamma f_t + F'(f) + uG'(f) \tag{4}$$
$$p = \nu \nabla f \tag{5}$$

where γ and ν are two positive constants while $F'(f)$ and $G'(f)$ are the derivatives of two functions $F(f)$ and $G(f)$ such that $F(f)$ enters as a potential in the free energy and $(u - G(f))$ is the entropy function. Finally, u is the parameter that controls the transition. In solid-fluid transition and in ferromagnetism u is the temperature, in superconductivity u is the velocity p_s, for turbulence phenomena u is the square of the velocity.

Therefore, in any of these phenomena the Ginzgurg-Landau equation assumes the form

$$\rho\left[\gamma f_t + F'(f) + uG'(f)\right] = \nu \nabla^2 f + \rho\sigma \ . \tag{6}$$

3 First order phase transition

The typical first order phase transition occurs when a solid phase (ice) melts to a liquid phase (water) and vice versa. In this section, by means of non-convex free energies, we shall prove that it is possible to use the equation (6) for modelling a general first order phase transitions. So we suggest for this phenomenon the following differential system

$$f_t = \frac{1}{\kappa^2}\nabla^2 f + f^2(1-f) - f(1-f)^2(1+\tau) \tag{7}$$

where $\tau = \frac{T-T_c}{T_c} > -1$ and c, k are positive coefficients.

Now we are in a position to prove that this model can be written in the form (??,??) with $u = 1+\tau$ and where

$$G(f) = \left(\frac{f^2}{2} - \frac{2}{3}f^3 + \frac{f^4}{4}\right) \quad , \qquad F(f) = (f^4/2 - f^3/3)$$

Moreover, this model represents a phase transition of first order. Indeed, when τ is not too far from the critical temperature $\tau = 0$, the energy

$$E(f) = F(f) + uG(f)$$

presents two local minima in $f = 0$, and $f = 1$.

Moreover, at the critical temperature $\tau_c = 0$, the latent heat l is given by

$$l = G(1) - G(0) = \left(\frac{f^2}{2} - \frac{2}{3}f^3 + \frac{f^4}{4}\right)_1 - \left(\frac{f^2}{2} - \frac{2}{3}f^3 + \frac{f^4}{4}\right)_0 \neq 0$$

Therefore the phase transition is of first order. Moreover, this model presents the advantage that the two local minima of the energy $E(f)$ are invariant for all $\tau \in (-1, \infty)$.

Now, we are in a position to prove a maximum theorem for the unknown f.

Theorem 1. *If for $t = 0$ we have $0 \leq f(0) \leq 1$, then $0 \leq f(t) \leq 1$ for any $t > 0$.*

Proof. The equation (??) can be rewritten in the new form

$$f_t = \frac{1}{\kappa}\nabla^2 f + f(1-f)\left[f + (f-1)(1+\tau)\right] \tag{8}$$

Now, consider the function $f_- = \max(-f, 0)$, whose derivative is given by

$$(f_-)_t = \begin{cases} 0, & f > 0 \\ -f_t, & f < 0 \end{cases}$$

From (8) we obtain

$$f_- f_t = \frac{1}{\kappa} f_- \nabla^2 f + f_- f(1-f)\left[f + (f-1)(1+\tau)\right] \tag{9}$$

where in particular we have $f_-(f_-)_t = -f_- f_t$. Therefore, from (9), we get

$$\frac{1}{2}\int_\Omega (f_-^2)_t dx \leq -\frac{1}{\kappa^2}\int_\Omega (\nabla f_-)^2 dx - \int_\Omega f_- f^2(1-f)dx$$

from which, by using the initial condition $0 \leq f(0) \leq 1$, we have

$$\frac{1}{2}\int_\Omega f_-^2(t)dx + \int_0^t \int_\Omega [\frac{1}{\kappa^2}(\nabla f_-)^2 + f_- f^2(1-f)]dx \leq 0$$

The inequality $f_-(1-f) \geq 0$ implies

$$f_-(t) = 0, \quad \text{for all } t > 0$$

whence $f \geq 0$, for $t > 0$.

Now, we prove the second part of the theorem, namely $f(t) \leq 1$, for $t > 0$. To this propose let us multiply (8) by $(f-1)_+ = \max(f-1, 0)$

$$(f-1)_+ f_t \leq \frac{1}{\kappa^2}(f-1)_+ \nabla^2 f - f^2(1-f)_+^2$$

Finally, by integrating on Ω

$$\frac{1}{2}\int_\Omega ((f-1)_+^2)_t dx \leq -\frac{1}{\kappa^2}\int_\Omega (\nabla(f-1)_+)^2 dx - \int_\Omega f^2(1-f)_+^2 dx$$

From this inequality and by means of the assumption of condition $f(0) \leq 1$, necessarily we may state

$$(f-1)_+(t) = 0, \quad \text{for all } t > 0$$

which is the thesis of the theorem.

4 Water-vapor phase transition

For the water-vapor phase transition, the order parameter equation is given by

$$\rho\gamma\dot{f} = \frac{\varkappa_0}{\rho}\nabla\cdot\rho\nabla f - \rho\theta_c F'(f) - \rho(\theta + \alpha(\theta,\theta_c)(p - p_0))G'(f) \tag{10}$$

where θ_c is the *critical temperature*, while p_0 is a suitable constant.

For this model we suggest that

$$\alpha(\theta,\theta_c) = \begin{cases} \alpha_0\frac{\theta_c - \theta}{\theta} \ , & for\ \theta < \theta_c \\ 0 \ , & for\ \theta \geq \theta_c \end{cases} \tag{11}$$

the coefficient $\alpha_0 > 0$ is related with the angular coefficient of the line (Fig.3) between the water ($f = 0$) and vapor ($f = 1$). Moreover, in (10) we have the same functions F and G given in (??).

We consider the density composed of two terms, ρ_0 and $\rho_1$which are related with the pressure and with the structure order, respectively. Moreover, we suppose the thermodynamic potential ρ_0,ψ,η function of the state σ of the system, given by the triple $\sigma = (\rho_0, f, \theta, \nabla f)$.Now the material will be supposed compressible. Hence the pressure will not be a indeterminate quantity, but a function of the density ρ_0 and temperature θ.

The First law is always represented by

$$\rho\dot{e} = \rho h + \mathcal{P}_i^f + \mathcal{P}_i^v \tag{12}$$

where the two power are now given by the following relations

$$\begin{aligned}\mathcal{P}_i^f &= \rho\Big[\gamma\dot{f}^2 + \varkappa_0 \nabla f \cdot \nabla\dot{f} + \theta_c \dot{F}(f) \\ &\quad +(\theta + \alpha(\theta,\theta_c)(p - p_0))\dot{G}(f)\Big]\end{aligned} \tag{13}$$

$$\mathcal{P}_i^v = (p - p_0)\frac{\dot{\rho}}{\rho} + \nu(f)\nabla\mathbf{v}\cdot\nabla\mathbf{v} \tag{14}$$

In this framework the Second law (??) takes the form

$$\begin{aligned}&\rho(\psi_\theta + \eta)\dot{\theta} - \rho\gamma f^2 + \rho(\psi_f - \theta_c F' - \theta G')f + \rho\psi_{\rho_0}\dot{\rho}_0 + \nu(f)\nabla\mathbf{v}\cdot\nabla\mathbf{v} \\ &-\rho(p - p_0)(\frac{\dot{\rho}}{\rho^2} + \alpha(\theta,\theta_c)\dot{G}(f)) + \rho(\psi_{\nabla f} - \varkappa_0\nabla f)\cdot\nabla f - \frac{\kappa_0}{\theta}|\nabla\theta|^2 \leq 0.\end{aligned} \tag{15}$$

Similarly to the ice-water phase transition, we suppose for the density an analogous constitutive equation

$$\frac{\dot{\rho}}{\rho^2} - \alpha(\theta,\theta_c)\dot{G}(f) = \frac{\dot{\rho}_0}{\rho_0^2} \tag{16}$$

Since, the temperature is constant during the transition, the equation (16) can be written

$$\frac{d}{dt}(\frac{1}{\rho} - \alpha(\theta,\theta_c)G(f)) = \frac{d}{dt}\frac{1}{\rho_0} \tag{17}$$

on any isothermal process. Consequently, on such processes we have the following constrain on ρ, ρ_0, θ and f

$$\frac{1}{\rho} - \alpha(\theta,\theta_c)G(f) = \frac{1}{\rho_0}$$

Therefore, also during the water-vapor transition, the density is expressed by means of two elements ρ_0 and ρ_1 as follows

$$\frac{1}{\rho_0} + \frac{1}{\rho_1} = \frac{1}{\rho}$$

where

$$\frac{1}{\rho_1} = \alpha(\theta,\theta_c)G(f). \tag{18}$$

In this framework, the pressure is related with the density ρ_0 and with the temperature by the constitutive equation

$$p = p(\rho_0, \theta) \tag{19}$$

while, the density ρ_1 , which does not enter in (19), is a function of the phase and temperature only.

We observe that the length $\left(\frac{1}{\rho_1}\right)_{\max}$ of the horizontal line of the Andrews diagram will be a function of the temperature. Of course, when $\theta = \theta_c$ the horizontal line is equal to zero. Moreover, $\left(\frac{1}{\rho_1}\right)_{\max}$ increases as the temperature decreases; in fact, using (18) we have

$$\left(\frac{1}{\rho_1}\right)_{\max} = \frac{\alpha_0}{12}\left(\frac{\theta_c}{\theta} - 1\right) \quad for \ \ \theta < \theta_c$$

Moreover, because the free energy ψ and the entropy η are assumed to be functions of the state $\sigma = (\rho_0, f, \theta, \nabla f)$,then the inequality (15) take the new form

$$\begin{aligned} &\rho(\psi_\theta + \eta)\dot\theta - \rho\gamma f^2 + \rho(\psi_f - \theta_c F' - \theta G')f \\ &-\nu(f)\nabla\mathbf{v}\cdot\nabla\mathbf{v} + \left[\rho\psi_{\rho_0} - \rho\frac{p(\rho_0,\theta) - p_0}{\rho_0^2}\right]\dot\rho_0 \\ &+\rho(\psi_{\nabla f} - \varkappa_0\nabla f)\cdot\nabla f - \frac{\kappa_0}{\theta}|\nabla\theta|^2 \le 0. \end{aligned} \tag{20}$$

Therefore, from (20) by virtue of the arbitrariness of $(f, \dot\theta, \dot\rho_0, \nabla f)$, it follows that

$$\eta = -\psi_\theta, \quad \psi_{\mathrm{f}} - \theta_c \mathrm{F}' - \theta \mathrm{G}' = 0 \ , \tag{21}$$

$$\psi_{\nabla f} - \varkappa_0 \nabla f = 0, \tag{22}$$

$$\psi_{\rho_0} - \frac{p(\rho_0,\theta) - p_0}{\rho_0^2} = 0$$

In order to obtain the Andrews diagram it is convenient to choose for ψ a function such that

$$\psi_{\rho_0} = \frac{\beta(\theta)}{\rho_0} \ , \quad \beta(\theta) > 0 \tag{23}$$

from which, if in particular $\beta = nR\theta$, we have the classical equation of a perfect gas

$$p(\rho_0,\theta) - p_0 = \rho_0^2\psi_{\rho_0} = \beta(\theta)\rho_0 = nR\rho_0\theta \tag{24}$$

where R is a universal constant and n denotes the number of moles of the gas.

Thus, the corresponding is

$$\begin{aligned} \psi \quad = \quad & \int e_\theta(\theta)d\theta - \theta\int\frac{e_\theta}{\theta}d\theta \\ & +\theta_c F(f) + \theta G(f) + \frac{\varkappa_0}{2}(\nabla f)^2 + nR\theta\log\rho_0 \end{aligned} \tag{25}$$

from which we have the entropy function

$$\eta = \int\frac{e_\theta}{\theta}d\theta - G(f) - nR\log\rho_0 \tag{26}$$

and the following expression for the internal energy

$$e = \theta_c F(f) + \frac{\varkappa_0}{2}(\nabla f)^2 + \int e_\theta d\theta$$

Finally, the inequality (20) reduces to

$$-\gamma\rho f^2 - \kappa_0|\nabla\theta|^2 \leq 0. \tag{27}$$

In this framework, the differential problem is now defined by the following system

$$\gamma f = \frac{\varkappa_0}{\rho}\nabla\cdot\rho\nabla f - \theta_c F'(f) - (\theta + \alpha(\theta,\theta_c)(p-p_0))G'(f) \tag{28}$$

$$\rho\dot{\mathbf{v}} = -\nabla p(\rho_0,\theta) + \rho\mathbf{b} \tag{29}$$

$$\dot{\rho} + \rho\nabla\cdot\mathbf{v} = 0 \tag{30}$$

$$\frac{\dot{\rho}}{\rho^2} = \frac{\dot{\rho}_0}{\rho_0^2} - \alpha(\theta,\theta_c)\dot{G}(f) \tag{31}$$

$$\rho(\frac{e_\theta}{\theta}\dot{\theta} - \dot{G}(f) - nR(\log\rho_0)^{\cdot}) = \kappa_0\nabla^2\log\theta + \frac{\rho r}{\theta} \tag{32}$$

Multiplying (28) by ρf, (32) by θ we have the energy balance

$$\begin{aligned} &\rho((\frac{\mathbf{v}^2}{2})^{\cdot} + \dot{e}\,) + \rho\gamma f^2 + \kappa_0\frac{1}{\theta}(\nabla\theta)^2 \\ = \;& \varkappa_0\nabla\cdot(\rho f\nabla f) + \nabla\cdot(\kappa_0\theta\Delta\log\theta) + \rho\mathbf{b}\cdot\mathbf{v} + \rho r \end{aligned} \tag{33}$$

Hence, by an integration on the domain Ω we obtain the global form of energy theorem

$$\frac{d}{dt}\int_\Omega \left[\rho(\frac{\mathbf{v}^2}{2} + \dot{e})\right] dx \leq \tag{34}$$

$$\leq \int_{\partial\Omega}(f\nabla f - \kappa_0\theta\nabla\log\theta)\cdot\mathbf{n}dS + \int_\Omega \rho(\mathbf{b}\cdot\mathbf{v} + r)dx$$

Finally, in order to obtain the Andrews diagram, we study in the plain (p,ρ) the function, which we obtain from (24) and the equation

$$\theta - \alpha(\theta,\theta_c)(p-p_0) = \theta_c \tag{35}$$

from which for $\theta < \theta_c$ we have

$$(\theta-\theta_c)(1 - \frac{\alpha_0}{\theta}(p-p_0) = 0$$

that describes the line which demarcates the liquid-vapor phase transition. Then for $\theta < \theta_c$ we have

$$p = \frac{\theta}{\alpha_0} + p_0 \tag{36}$$

or by means of (24)

$$p(1 - \frac{1}{\alpha_0 nR\rho_0}) = p_0 \tag{37}$$

hence

$$p = \frac{p_0\alpha_0 nR\rho_0}{\alpha_0 nR\rho_0 - 1} = \frac{p_0\alpha_0 nR}{\alpha_0 nR - \frac{1}{\rho_0}} \tag{38}$$

Moreover, the critical value p_c and θ_c are jointed by

$$p_c = \frac{\theta_c}{\alpha_0} + p_0 \tag{39}$$

Therefore, from (38) and (39) we obtain

$$\frac{\theta_c}{\alpha_0} + p_0 = \frac{p_0 \alpha_0 n R \rho_c}{\alpha_0 n R \rho_c - 1}$$

from which $\frac{1}{\rho_c} < \alpha_0 n R$. Then, the graphic $p = p(v, \theta, f)$ $(v = \frac{1}{\rho})$ is able to describe the Andrews diagram, and it will be of the type

$$p - p_0 = nR\,(v - \alpha(\theta, \theta_c)G(f))^{-1}\,\theta \tag{40}$$

with

$$v - \alpha(\theta, \theta_c)G(f) = v_0 \tag{41}$$

References

[1] J. Bardeen, Theory of superconductivity. In Handbuch der Physik XV (Edited by S. Flügge), (Springer, 1956) pp 274–369.

[2] G. Du, M.D. Gunzburger and J.S. Peterson, Analysis and approximation of the Ginzburg-Landau model of superconductivity. SIAM Rev. 34, (1992) 54-81.

[3] Fabrizio M., Ginzburg-Landau equations and first and second order phase transitions. Int. J. Engng Sci. 44 (2006), 529-539.

[4] M. Fabrizio, G. Gentili and B. Lazzari, A nonlocal phenomenological theory in superconductivity, Math. Models Methods Appl. Sci. 7, (1997) 345–362.

[5] M. Frémond, Non-smooth thermomechanics. Springer-Verlag, Berlin, 2002, xvi+480.

[6] E. Fried and M. Gurtin, Continuum theory of thermally induced phase transitions based on a order parameter, Physica D 68, (1993) 326-343.

[7] V. L. Ginzburg and L. D. Landau, On the theory of superconductivity, Zh. Eksp. Teor. Fiz. 20, (1950) 1064–1082.

[8] L. P. Gor'kov and G.M. Éliashberg, Generalization of the Ginzburg-Landau equations for nonstationary problems in the case of alloys with paramagnetic impurities, Soviet Phys. JETP 27, (1968) 328–334.

[9] L. D. Landau, Unbranched model of intermediate state, Phys. Z. Sowjet. 11, (1937) 129–138.

[10] L. D. Landau, On the theory of superfluidity of helium II, J. Phys. USSR. 5, (1941) 71-77.

Sui Problemi al Contorno Mobile

J.N. Flavin

Mathematical Physics Department,
National University of Ireland,
Galway, Ireland.

Dedicato a Brian Straughan

Abstract

The context is that of the heat equation in a straight line region with one fixed end, while the other has an assigned motion. A zero, or zero flux, condtion is imposed on the fixed end, while on the moving end the temperature is assigned as a function of time. A functional is defined in this context and an interesting identity is given for this. In the context of an expanding region, an upper bound is given for $F(t)$ in terms of data (initial conditions included). It is proved that the upper bound is asymptotically *sharp*. Generalizations are discussed.

Diffusione Unidimensionale con un Contorno Mobile.

Si considera una regione rettilinea con un contorno fisso $x = 0$ e un contorno mobile $x = L(t)$, dove $L(t)$ è una funzione liscia assegnata. Si considerano funzioni liscie $u(x,t)$ che verificano l'equazione di diffusione

$$u_t = u_{xx} \tag{1}$$

e le condizioni al contorno

$$u = 0 \quad \text{o} \quad \mathrm{u_x} = 0, \quad \mathrm{x} = 0;$$

$$u = \phi(t), \qquad x = L(t), \tag{2}$$

dove $\phi(t)$ è una funzione liscia assegnata. Si suppone anche che $u(x,0)$ sia assegnato. In questo contesto, x *indice* significa differenziazione partiale rispetto a x etc., e un punto *sovrapposto* significa differenziazione ordinaria rispetto a t.

Si definisce il funzionale

$$F(t) = \int_0^{L(t)} u_x^2 dx \tag{3}$$

lungo le soluzioni di (1),(2).

Theorem 1. *Il funzionale (3) verifica l'identità*

$$\frac{dF}{dt} = -2\int_0^{L(t)} u_x^2 dx + \frac{\dot{\phi}^2(t) - \{u_t(L,t)\}^2}{\dot{L}},\ \dot{L} \neq 0. \tag{4}$$

Remark 1. *Si deve notare la compattezza inaspettata del termine al contorno nell'identità (4).*

Si proverà una versione più generale dell'identità (4) in un altro articolo [1]; cf.[2]. Nel contesto di una regione in espansione, $\dot{L} > 0$, si può dedurre da (4)

$$\frac{dF}{dt} \leq \dot{\phi}^2(t)/\dot{L}(t).$$

Integrando, si ottiene

Theorem 2. *Nel contesto di una regione in espansione, $\dot{L} > 0$, si ha*

$$F(t) \leq \int_0^t \left\{\dot{\phi}^2(\tau)/\dot{L}(\tau)\right\} d\tau; \tag{5}$$

(5) è anche valido se si ha $\dot{L} \geq 0$, e esiste l'integrale

$$\int_0^t d\tau/\dot{L}(\tau) < \infty.$$

Theorem 3. *La disuguaglianza (5) è* sharp *asintoticamente.*

La dimostrazione del Teorema usa un approccio inverso.
Si considera, per semplicità (e per primo):

(a) la condizione al contorno

$$u_x(0,t) = 0, \tag{6}$$

(b) la soluzione polinomiale di (1)

$$u = x^2 + 2t, \tag{7}$$

(c) il moto del contorno

$$L(t) = 1 + t. \tag{8}$$

Ovviamente (7) verifica (6). Le condizioni (7), (8) forniscono i *dati*:

$$\phi(t) = (1+t)^2, \tag{9}$$

e

$$u(x,0) = x^2 \tag{10}$$

I dati (1), (2), (6), (8)-(10) forniscono la soluzione unica (7).

Indicando il maggiorante di $F(t)$ con $\overline{F}(t)$ [5], si calcola direttamente

$$\overline{F}(t)/F(t) = \left\{(t+2)^3 - 7\right\}/(t+1)^3 \sim 1, t \to \infty. \tag{11}$$

Quindi si stabilisce il Teorema 3. Se si ha la condizione al contorno $u(0,t) = 0$, invece di (6), la prova è simile ma i dettagli sono diversi.

Si può generalizzare il Teorema 3 notevolmente: nel contesto della condizione al contorno (6), si suppongono

(a)

$$u(x,t) = \sum_{m=0}^{n} a_{2m} P_{2m}(x,t) \tag{12}$$

dove

(i) a_{2m} sono costanti, $a_{2n} \neq 0$,

(ii) P_{2m} significa la soluzione polinomiale dell'equazione (1):

$$P_{2m}(x,t) = \sum_{k=0}^{m} \frac{x^{2m-2k} t^k}{(2m-2k)!k!}$$

(b) il moto (di espansione)

$$L(t) = \sum_{v=0}^{p} b_r t^r \tag{13}$$

dove le b sono costanti non-negative.

$$b_0 \neq 0, b_1 \neq 0, b_p \neq 0,$$

p essendo un intero positivo.

Le condizioni (12),(13) forniscono i *dati* $u(x,0), \phi(t)$ dove

$$\phi(t) = \sum_{m=0}^{n} a_{2m} P_{2m}(L,t) \tag{14}$$

un'espressione abbastanza generale. Si può provare questa generalizzazione del corollario, usando l'identità (4), integrata; questa generalizzazione fornisce non solamente il limite asintotico ma anche il tasso di convergenza:

Theorem 4. *Nel contesto dei* dati *forniti da (6), (12),(13), si ha*

$$\overline{F}(t)/F(t) = 1 + O(t^{-2p+1}) \tag{15}$$

quando $t \longrightarrow \infty$, in cui si usa la notazione usuale per l'ordine di grandezza.

Delle Osservazioni

(i) Si può provare che i risultati asintotici (11), (15) sono independenti dal valore $u(x,0)$ all'inizio: cioè si ottengono gli stessi risultati asintotici quando si cambia la condizione iniziale, ma si mantengono tutti gli altri *dati*.

(ii) C'è un senso in cui si può approssimare qualsiasi soluzione liscia di (1), pari in x con accuratezza arbitraria, con un'espressione del tipo (12), e.g. [3]: quindi la rappresentazione (12) è abbastanza generale.

(iii) Nel contesto dei Teoremi 3, 4 si può sostituire la condizione al contorno $u_x(0,t) = 0$ con la condizione $u(0,t) = 0$. Gli stessi risultati sono validi, ma si devone cambiare i dettagli delle prove.

(iv) in contesto generale, dove $u_x(0,t) = 0$, si può ottenere un maggiorante per

$$|u(x,t) - \phi(t)|$$

in termini dei *dati*, usando (5) e la disuguaglianaza di Schwarz. Se si ha $u(0,t) = 0$, si possono dedurre altre stime puntuali usando gli stessi metodi generali.

(v) Si possono generalizzare i risultati di questo articolo. Si possono ottenere dei risultati dello stesso tipo

(a) per lo stesso contesto nella regione con due contorni mobili,

(b) per un sistema di equazioni con due contorni mobili.

(vi) Si possono usare metodi simili per analizzare certe equazioni nonlineari, e.g. [4].

(vii) A.S. Fokas e i suoi coautori hanno pubblicato alcuni articoli su problemi con contorni mobili, usando metodi diversi, e.g. [5] –[7].

References

[1] J.N. Flavin, articolo non ancora publicato.

[2] J.N. Flavin and S. Rionero, *Qualitative Estimates for Partial Differential Equations,An Introduction,* CRC Press, 1995.

[3] D.L. Colton, *Analytic Theory of Partial Differential Equations*, Monographs and Studies in Mathematics **8**,Pitman, 1980.

[4] J.N. Flavin and C. Rogers, Upper estimates for a Resonant Nonlinear Schrödinger Equation, *Studies in Applied Mathematics* (to appear).

[5] A.S. Fokas and P.F. Schultz, The Long-Time Asymptotics of Moving Boundary Problems Using and Ehrenpreis -Type Representation and its Riemann-Hilbert Nonlinearisation. *Communications of Pure and Applied Maths.* **LVI,** 1-40 (2002).

[6] A.S. Fokas and B. Pelloni, Generalised Dirichlet to Neumann Map for Moving Boundary Value Problems, *J. Math. Phys.* **48**, 013502 (2007).

[7] S. de Lillo and A.S. Fokas, The Dirichlet to Neumann Map for the Heat Equation on a Moving Boundary, *Inverse Prob.*, **23**, 1699-1710 (2007).

A Phase-Field Model for Liquid-Vapor Transitions induced by Temperature and Pressure

*A. Berti, C. Giorgi**

Dipartimento di Matematica
Università di Brescia,
via Valotti 9, 25133 Brescia, Italy

Abstract

We scrutinize the equations ruling the liquid-vapor phase transition. Starting from the macroscopic state equations of the vapor and the liquid pure phases, we propose a phase-field model which is thermodynamically consistent and accounts for both temperature and pressure variations during the evaporation process. In particular, we are able to determine the explicit expression of the Gibbs free enthalpy. As a consequence, the Clausius-Clapeyron formula is obtained and the customary form of the vapor pressure curve is recovered.

Keywords: Phase field, liquid-vapor transition, vapor pressure curve, thermodynamics.

1 Introduction

We propose a non isothermal, non isobaric phase-field model, describing the first order transitions between two different states of the matter, for instance solid-liquid or liquid-vapor. Since different phases are endowed with different structures or symmetries, for a large number of materials we are allowed to introduce an order parameter, χ, summarizing the structure order of the matter. According to a customary setting, the scalar parameter χ takes its values in $\{0,1\}$ and characterizes the *amount of order* of the internal structure, so that $\chi = 0$ corresponds to the less-ordered one. An example occurs in a solid-fluid (liquid-vapor) phase transition, when a solid (liquid) phase melts (vaporizes) to a liquid (gaseous) phase. In the particular case of water, the order parameter χ is identified with the ice (water) concentration, so that $\chi = 0$ for water (vapor), the less ordered phase, and $\chi = 1$ for ice (water).

Let Ω be the three-dimensional domain occupied by two phases of the same particle species. At the macroscopic scale, a sharp interface (of fractal dimension, maybe) separates two regions occupied by different phases. As a consequence, the order parameter is piecewise constant in Ω and this approach leads to well-known free-boundary

*Email: giorgi@ing.unibs.it

problems where the main effort, from the mathematical point of view, is to predict the localization and the evolution of the phase interface. Unfortunately, numerical simulations are very hard in this framework. In order to avoid this difficulty, in recent years many papers adopted a different approach, named *phase-field* modeling (see, for instance, [1, 3, 4, 8, 13, 14] and reference therein). It lies on the level of the mesoscopic scale, in that phase interfaces are regarded as thin layers rather than sharp surfaces. This feature is captured if we introduce a regular scalar field

$$\varphi(x,t) : \Omega \times [0,T] \to [0,1]$$

whose values overlap the order parameter χ throughout the domain Ω, except for a very thin region around the phase interfaces where $\nabla\varphi$ is noticeably different from zero. A specific contribution, which is named *interface energy* and depends on the width ϵ of that region, must be added to the free energy expression. As proved in [11], phase-field models of first-order phase transitions are more tractable from the numerical point of view and in the limit $\epsilon \to 0$ they approach reasonably well free boundary problems for melting and nucleation processes.

Usually, non isothermal transitions are depicted as temperature induced phenomena at constant pressure, say p^* [1, 8, 13]. If this is the case, the (absolute) temperature field

$$\theta(x,t) : \Omega \times [0,T] \to \mathbb{R}^+$$

must be added to the phase-field in order to describe the evolution of the system. Then, a temperature value θ^*, matching the fixed pressure p^*, is introduced into the model in such a way that the homogeneous stationary solutions $\varphi(x,t) = 0$ and $\varphi(x,t) = 1$, $(x,t) \in \Omega \times [0,T]$, are both stable when $\theta(x,t) = \theta^*$. In addition, it is usually assumed that the less-ordered phase $\varphi = 0$ is stable when $\theta > \theta^*$ and, on the contrary, the phase $\varphi = 1$ is stable when $\theta < \theta^*$. Such a behavior is accounted for by means of the special form of the free energy dependence on θ and φ (see, for instance, [5, 6]). Within this framework, all transition processes are considered at the same constant pressure in $x \in \Omega$, and this is consistent with the observation that during first-order phase changes *both temperature and pressure keep constant values* (θ^*, p^*). This occurrence introduce a natural pairing between the temperature θ^* and the matching value of the pressure, p^*.

1.1 The phase diagram

In order to take into account the effects due to pressure variation, we need to introduce the pressure field

$$p(x,t) : \Omega \times [0,T] \to \mathbb{R}.$$

In general, a phase transition occurs in (x,t) when $(\theta(x,t), p(x,t))$ crosses the point (θ^*, p^*). Accounting for experimental data, for any specified pair of phases (*i.e.* solid–liquid, liquid–vapor, solid–vapor) this procedure yields a different curve in the (θ, p)–plane. The resulting curves cross in the so-called *triple point*, (θ_T, p_T), and split the (θ, p)–plane into three regions (see fig.1). In the case of water, $\theta_T = 0.0098°C$ and $p_T = 0.00603$ atm. Superheating and undercooling phenomena prove that each branch of the diagram does not separate states corresponding to different phases, but it edges two regions where different phases are stable [10].

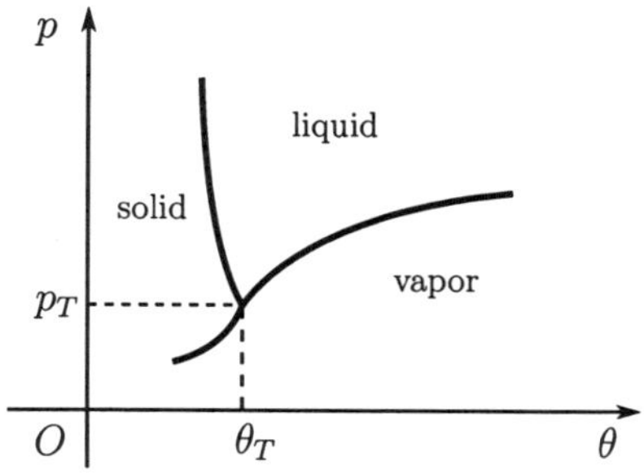

Figure 1: *The phase diagram in the θ–p plane*

The curve between the regions of solid and liquid phase stability is called *melting pressure*, and is represented by the function $p^* = \pi_m(\theta^*)$. On the other hand, vapor and liquid phase stability regions are separated by the so-called *vapor pressure* curve given by $p^* = \pi_v(\theta^*)$. These curves are not to be mistaken for the customary functions of state at the liquid and vapor phases, that reads respectively

$$p = p_\ell(\rho,\theta)\,, \qquad p = p_v(\rho,\theta). \tag{1}$$

It is worth noting that π_m and π_v cross at the triple point, namely

$$\pi_v(\theta_T) = \pi_m(\theta_T) = p_T.$$

In particular, the pressure curve of vaporization ends abruptly in a state (θ_c, p_c), named *critical point.* It is defined as the state above which a phase boundary between liquid and vapor ceases to exist (see fig.2). In water, it occurs at around 374 $°C$ and 218 atm.

It is an easy matter to prove that π_m and π_v must obey the *Clausius–Clapeyron equation* (see, for instance, [7] p.77 and [10] Example 3.1). In particular, the *vapor pressure* π_v is the solution of the first order differential equation

$$\pi_v'(\theta^*) = \frac{L_v\,\rho_v\rho_\ell}{\theta^*(\rho_\ell - \rho_v)}, \tag{2}$$

where L_v is the *evaporation latent heat* and ρ_ℓ, ρ_v are respectively the liquid and vapor density functions related to (1). Since the right-hand side of (2) is evaluated at a point of the vapor pressure curve, it can in general be a complicated function of $(\theta^*, \pi_v(\theta^*))$. Far from the critical point (θ_c, p_c) the approximation $\rho_v \ll \rho_\ell$ is allowed and (2) reduces to $\pi_v'(\theta^*) \approx L_v\rho_v/\theta^*$. Finally, assuming that L_v is constant and ρ_v obeys the equation of state of a perfect gas (5), we obtain

$$\pi_v(\theta^*) = p_c \exp\Big(-\frac{L_v}{k\,\theta^*}\Big), \qquad \theta_T < \theta^* \ll \theta_c, \tag{3}$$

where k is the Boltzmann's constant. Of course, this expression of π_v represents a rough approximation of the real curve (see fig.2), in that $p_c = \lim\limits_{\theta^*\to+\infty} \pi_v(\theta^*)$ implies $\theta_c = +\infty$.

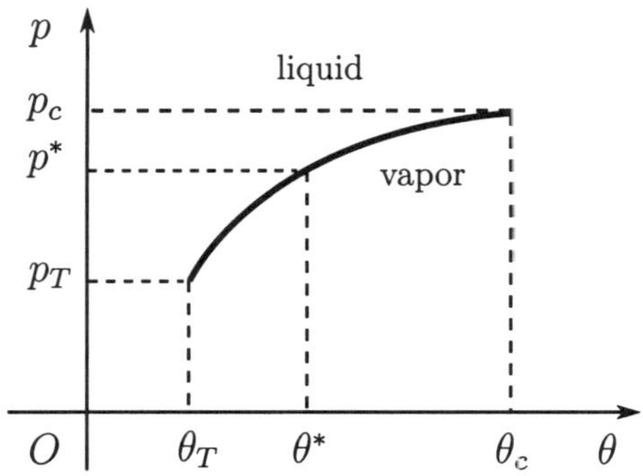

Figure 2: *The vapor pressure curve*

Taking into account that the triple point lies on this curve, we obtain the following relation

$$p_c = p_T \exp\left(\frac{L_v}{k\,\theta_T}\right).$$

In a similar way the *melting pressure* curve obeys

$$\pi_m'(\theta^*) = \frac{L_m\,\rho_l\rho_s}{\theta^*(\rho_s - \rho_\ell)} = \frac{L_m\overline{\rho}}{\theta^*}, \qquad \overline{\rho} = \frac{\rho_l\rho_s}{\rho_s - \rho_\ell}$$

where ρ_s and ρ_ℓ are respectively the solid and liquid densities ($\rho_s \approx \rho_\ell$), and L_m is the *melting latent heat.* Note that $\overline{\rho}$ is approximately independent of p and θ, and is negative when $\rho_s < \rho_\ell$, but it is positive if $\rho_s > \rho_\ell$. In the former case (for instance, in the ice–water transition), the slope of this curve is negative and π_m is defined on $(0, \theta_T]$. In both occurrences, the smallness of the term $|\rho_s - \rho_\ell|$ yields a very large slope of the melting pressure curve so that in a rough approximation scheme it reduces to the line $\theta = \theta_T$.

In all kind of transitions (solid-liquid or liquid-vapor), π_m and π_v are monotone increasing functions, with the exception of the water where π_m is decreasing. In any case we can introduce their inverse functions, namely $\tau_m = \pi_m^{-1}$ and $\tau_v = \pi_v^{-1}$, such that

$$\theta^* = \tau_m(p^*)\,, \qquad \theta^* = \tau_v(p^*).$$

1.2 The energy–temperature and Andrews density–pressure diagrams

Henceforth, we restrict our attention to phase transitions between liquid and vapor. If we consider the internal energy density ε, both in the vapor and in the liquid phase it is a function of the temperature θ, only. Letting ε_0 the internal energy in the vapor and ε_1 in the liquid, it is customary to assume that

$$\varepsilon_0(\theta) = c_0\theta, \qquad \varepsilon_1(\theta) = c_1\theta - L, \tag{4}$$

where L is constant and strictly positive and the rates c_0 and c_1 denotes the vapor and liquid specific heat at constant volume, respectively. When water is concerned, the specific heat of the vapor is greater than that of the liquid, and $c_0 < c_1$, accordingly. The positive jump of the energy at the critical temperature θ^*, namely

$$\varepsilon_0(\theta^*) - \varepsilon_1(\theta^*) = [c_0 - c_1]\theta^* + L,$$

represents the *latent heat* L_v absorbed during the evaporation process and is related to the length of the horizontal segment shown in figure 3 (on the left the liquid energy, $\theta < \theta^*$, on the right the vapor energy, $\theta > \theta^*$). If $c_0 = c_1$, then $L_v = L$.

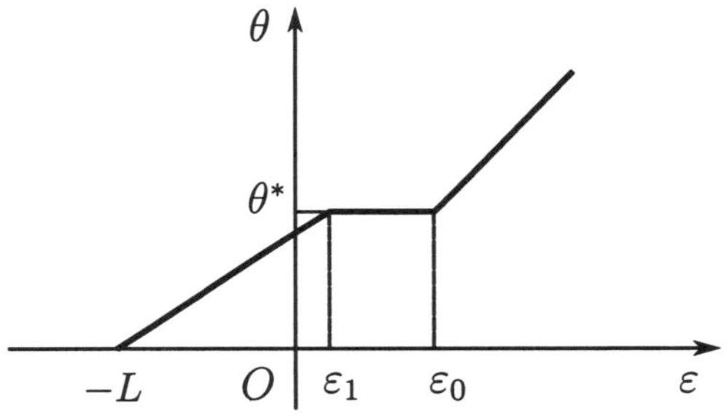

Figure 3: *The energy-temperature diagram.*

Now, our aim is to exploit the similarity between this diagram and the Andrews diagram (see fig.4) in order to represent the different equations of state (involving pressure, temperature and mass density) in the liquid and vapor phases. The Andrews diagram is obtained by applying the *Maxwell construction* to the isotherms of the van der Waals' equation (see, for instance, [10] p. 67). It mainly differs from the energy-temperature diagram by the presence of three variables rather than two. In fact, in the density–pressure plane we have a family of curves depending on the parameter θ (isothermal curves).

Let ρ_0 and ρ_1 be the density of the vapor and the liquid, respectively. We denote by v_0, v_1 the corresponding specific volumes, i.e. $v_0 = 1/\rho_0$ and $v_1 = 1/\rho_1$. As observed in the previous subsection, when $\theta \ll \theta_c$ then $\rho_0 \ll \rho_1$ and $v_0 \gg v_1$. It is quite reasonable to suppose that the vapor behaves like a perfect gas, so that

$$p_0 v_0 = k\theta\,.$$

This equation looks like a constitutive (state) equation for p_0 at fixed temperature, namely

$$p_0 = k\theta\,\rho_0. \tag{5}$$

Accordingly, we assume a similar constitutive law in the liquid phase

$$p_1 = k\theta[1 - \hat{\nu}(\theta)]\rho_1, \tag{6}$$

where $\hat{\nu}$ is a decreasing function of θ such that $\hat{\nu}(\theta) \in [0, 1)$ and satisfies the condition

$$\hat{\nu}(\theta) = 0, \qquad\qquad \theta \geq \theta_c,$$

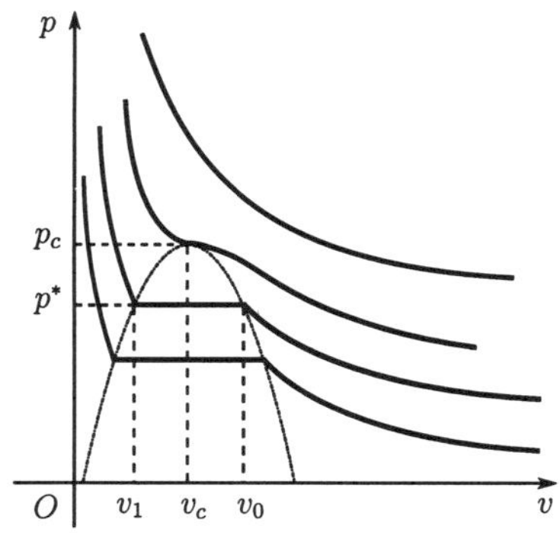

Figure 4: *The Andrews diagram in the v–p plane.*

where θ_c is the critical temperature. Indeed, when $\theta \geq \theta_c$ the vapor cannot be liquified at any pressure and hence the constitutive equation (5) should reduce to (6). For instance, we may choose

$$\hat{\nu}(\theta) = \begin{cases} \left(1 - \dfrac{\theta}{\theta_c}\right)\nu & \text{if} \quad 0 < \theta < \theta_c \\ 0 & \text{if} \quad \theta \geq \theta_c \end{cases}$$

where $0 < \nu < 1$, so that equation (6) takes the form

$$p_1 = \begin{cases} \dfrac{k}{\theta_c}\theta^2 \rho_1 & \theta < \theta_c, \\ k\theta\,\rho_1 & \theta \geq \theta_c. \end{cases}$$

The critical value v_c can be obtained from θ_c by virtue of the relation

$$v_c = \frac{k\theta_c}{p_c} = \frac{k\theta_c}{\pi_v(\theta_c)}.$$

We are finally in a position to evaluate the jump of density occurring at the phase change. To this end, we remember that both temperature and pressure keep constant across the transition process. Thus, we consider the isothermal curve $\theta = \theta^*$ in the Andrews diagram and, accounting for the knowledge of the vapor pressure function π_v, we are able to compute the value $p^* = \pi_v(\theta^*)$ at which the transition starts. During the evaporation process, the specific volume v (and hence the mass density ρ) moves from its value v_1 (ρ_1) in the liquid phase to the corresponding value v_0 (ρ_0) in the vapor phase, and both depend on θ^*, i.e.

$$v_0(\theta^*) = \frac{1}{\rho_0(\theta^*)} = \frac{k\theta^*}{\pi_v(\theta^*)}, \qquad v_1(\theta^*) = \frac{1}{\rho_1(\theta^*)} = \frac{k\theta^*[1 - \hat{\nu}(\theta^*)]}{\pi_v(\theta^*)}. \tag{7}$$

The assumption $\hat{\nu}(\theta^*) \in [0, 1)$ guarantees that $v_1(\theta^*) < v_0(\theta^*)$. Hence the positive jump of the specific volume during the transition is given by

$$v_0(\theta^*) - v_1(\theta^*) = \frac{k\hat{\nu}(\theta^*)\theta^*}{\pi_v(\theta^*)}.$$

This quantity decreases when temperature increases and vanishes as $\theta \to \theta_c$ (see fig.4).

2 Thermodynamics and phase-field equations

Although a rigorous derivation of the model will be addressed in a forthcoming paper, we sketch here the procedure.

We consider a bounded domain $\Omega \subset \mathbb{R}^3$ with smooth boundary $\partial\Omega$, which is completely filled by one chemical component (water, for instance). Our goal is to describe a transition between two phases (vapor and liquid) by means of a phase-field model. Accordingly, we suppose that there exists an interface with finite thickness separating the two pure phases and we assume that φ varies smoothly across the transition layer. The main challenge is to write the evolution equations for all the involved fields. To do this, besides the usual balance equations we need the relevant evolution equation for the phase-field φ and constitutive equations for the pressure p, the internal energy density ε and the heat flux vector $\boldsymbol{q}$. In particular, we look for a global description which is able to smoothly connect the extensive quantities across the interface layer.

We start by writing the balance equations of mass, momentum and energy

$$\dot{\rho} + \rho\nabla \cdot \boldsymbol{v} = 0 \tag{8}$$

$$\rho\dot{\boldsymbol{v}} = \nabla \cdot \boldsymbol{T} + \rho\boldsymbol{b} \tag{9}$$

$$\rho\dot{\varepsilon} = \boldsymbol{T} \cdot \nabla\boldsymbol{v} - \nabla \cdot \boldsymbol{q} + \rho r \tag{10}$$

where $\boldsymbol{v}$ is the velocity, $\boldsymbol{T}$ the Cauchy stress tensor, $\boldsymbol{b}$ the body force and r the external heat supply. Here and in the sequel a superposed dot denotes the total time derivative.

In order to account for non-local spatial effects, we introduce the following set of independent variables

$$\Sigma = (\rho, \theta, \varphi, \nabla\rho, \nabla\theta, \nabla\varphi, \Delta\rho, \Delta\theta, \Delta\varphi).$$

and we regard φ as a hidden variable whose evolution is described by

$$\dot{\varphi} = F(\Sigma). \tag{11}$$

Assuming a non-local form of the second law of thermodynamics ([9], [12]), we are lead to consider a regular field $\boldsymbol{k} = \boldsymbol{k}(x,t)$, named *entropy extra flux* such that $\boldsymbol{k} \cdot \mathbf{n} = 0$ on $\partial\Omega$, and

$$\rho\dot{\eta} \geq -\nabla \cdot \left(\frac{\boldsymbol{q}}{\theta} + \boldsymbol{k}\right) + \frac{\rho r}{\theta}. \tag{12}$$

where η is the entropy density. Let ψ be the Helmholtz free energy density. Taking into account the relation $\psi = e - \eta\theta$ and (10), we obtain the Clausius-Duhem inequality

$$-\rho(\dot{\psi} + \eta\dot{\theta}) + \boldsymbol{T} \cdot \nabla\boldsymbol{v} - \frac{1}{\theta}\boldsymbol{q} \cdot \nabla\theta + \theta\nabla \cdot \boldsymbol{k} \geq 0,$$

from which it can be proved (see [5]) that

$$\psi = \psi(\rho, \theta, \varphi, \nabla\varphi), \tag{13}$$

$$\partial_\theta \psi = -\eta \tag{14}$$

$$\boldsymbol{T} = -\rho^2 \partial_\rho \psi \boldsymbol{I} - \rho\, \mathrm{sym}(\nabla\varphi \otimes \partial_{\nabla\varphi}\psi) \tag{15}$$

$$\boldsymbol{k} = \frac{1}{\theta}\rho\, \partial_{\nabla\varphi}\psi\dot{\varphi} \tag{16}$$

$$\frac{1}{\theta^2}\boldsymbol{q}\cdot\nabla\theta + \left[\rho\frac{\partial_\varphi\psi}{\theta} - \nabla\cdot\left(\rho\frac{\partial_{\nabla\varphi}\psi}{\theta}\right)\right]\dot{\varphi} \le 0 \tag{17}$$

where ∂_ξ denotes the partial derivative with respect to ξ, $\boldsymbol{I}$ and $\mathrm{sym}\boldsymbol{A}$ stands for the identity tensor and the symmetric part of the tensor $\boldsymbol{A}$, respectively. In order to satisfy (17), it is natural to assume

$$\boldsymbol{q} = -\sigma\nabla\theta, \qquad \dot{\varphi} = -\alpha\left[\rho\frac{\partial_\varphi\psi}{\theta} - \nabla\cdot\left(\rho\frac{\partial_{\nabla\varphi}\psi}{\theta}\right)\right], \tag{18}$$

where σ, α are positive constants. In particular, the evolution of φ is ruled by the *rescaled* Helmholtz free energy density ψ/θ through $(18)_2$ (see, for a similar result, [1], [14]).

3 Free energy density

At this stage, we need informations about the Helmholtz free energy density ψ in order to uniquely and completely obtain the evolution equations of all unknown fields. To this end, we first assume here the customary expression introduced by Cahn-Hilliard (see [3])

$$\psi(\rho, \theta, \varphi, \nabla\varphi) = \Psi(\rho, \theta, \varphi) + \frac{\mu_0\theta}{2\rho}|\nabla\varphi|^2$$

where the last term accounts for the interface energy between the two phases and $\mu_0 > 0$. Then, starting form the definition of ψ and (14), we have

$$\psi - \theta\partial_\theta\psi = \Psi - \theta\partial_\theta\Psi = \varepsilon \tag{19}$$

Finally, observing that at uniform configurations (i.e. $\nabla\varphi = \boldsymbol{0}$) the Cauchy stress tensor $\boldsymbol{T}$ reduces to a normal pressure, namely

$$\boldsymbol{T}|_{\nabla\varphi=\boldsymbol{0}} = -p\boldsymbol{I},$$

from (15) we obtain

$$\partial_\rho\psi|_{\nabla\varphi=\boldsymbol{0}} = \partial_\rho\Psi = \frac{p}{\rho^2}. \tag{20}$$

As a consequence, if we give the constitutive equations for the pressure p and the internal energy ε inside the transition layer, then we can recover (to within a constant) the complete expression of the Helmholtz free energy. To this aim, we connect the *pure-phase* constitutive equations for p and ε (as devised in subsection 1.2) with a smooth function of the phase-field φ in such a way that θ and p keep constant (and hence continuous) across the transition layer, whereas ε and ρ suffer a jump discontinuity. More precisely, we choose

$$p(\rho,\theta,\varphi) = k\rho\,\theta[1-\nu g(\varphi)] \tag{21}$$
$$\varepsilon(\theta,\varphi) = c(\varphi)\theta + Lf(\varphi) \tag{22}$$

with f, g smooth functions satisfying the conditions

$$g(0) = f(0) = 0\,, \qquad g(1) = -f(1) = 1\,. \tag{23}$$

As expected, (4) and (5)-(6) follow from the evaluation of (21) and (22) at $\varphi = 0, 1$.

Here and in the sequel, we suppose that $\theta \ll \theta_c$ and, for sake of simplicity, we let $\hat{\nu} = \nu$ be constant. Moreover, since $v_1 \ll v_0$, we infer that ν lies between 1/2 and 1. Furthermore, we let

$$c(\varphi) = c_P - k[1-\nu g(\varphi)],$$

where $c_P = c_V + k$ is the specific heat at constant pressure in the vapor phase. Then, we have

$$c_0 = c(0) = c_V, \qquad c_1 = c(1) = c_V + k\nu,$$

and, according to previous statements, $c_0 < c_1$.

Remark 1. *Since during the transition $(p,\theta) = (p^*,\theta^*)$ are constant, the specific volume v in the transition layer depends on the order parameter φ, only. In particular,*

$$v(\varphi) = v_1 g(\varphi) + v_0[1-g(\varphi)]$$

which generalizes the usual linear relation $v(\varphi) = v_1\varphi + v_0(1-\varphi)$.

Proof. After substituting $\theta = \theta^*$ and $p = p^*$ into (21) we obtain

$$v(\varphi) = \frac{1}{\rho(\varphi)} = \frac{k\theta^*}{p^*}[1-\nu g(\varphi)] = \frac{k\theta^*}{p^*}(1-\nu)g(\varphi) + \frac{k\theta^*}{p^*}[1-g(\varphi)].$$

and accounting for (7) we have the required expression of the specific volume.

By exploiting (19)–(22) and previous arguments, we achieve (to within a constant)

$$\Psi(\rho,\theta,\varphi) = c_P\theta(1-\ln\theta) + Lf(\varphi) + k\theta[1-\nu g(\varphi)]\ln\frac{p}{p_c}\,, \tag{24}$$

with p given in (21) (see [2] for details). Notice that

$$\Psi(\rho,\theta,0) = c_P\theta(1-\ln\theta) + k\theta\ln\frac{p}{p_c}$$

meets the customary expression of the free energy for an ideal gas (see [7] eqn.120).

Remark 2. *The entropy of the vapor is greater than the entropy of the liquid provided that $p < p_c/e$ and $\theta < \tau_v(p_c/e)$.*

Proof. By taking (14) and (24) into account, we obtain

$$\eta_0(\rho,\theta,\varphi) = \eta|_{\nabla\varphi=\mathbf{0}} = c_P \ln\theta - k[1-\nu g(\varphi)]\left(\ln\frac{p}{p_c}+1\right).$$

Hence, the difference

$$\eta_0(\rho,\theta,0) - \eta_0(\rho,\theta,1) = -\nu k\left(\ln\frac{p}{p_c}+1\right) \tag{25}$$

is positive provided that $\ln p < \ln p_c - 1$, namely $p < p_c/e$. At a point $(\theta^*, \pi_v(\theta^*))$ of the vapor pressure curve this difference gives the transition entropy jump, and $\pi_v(\theta^*) < p_c/e$ implies $\theta^* < \tau_v(p_c/e) < \tau_v(p_c) = \theta_c$ which is consistent with the assumption $\theta \ll \theta_c$.

4 The vapor pressure curve

Since the temperature and the pressure are constant during the transition, we look for the expression of some potential depending on p and θ, rather than ρ and θ. Let

$$G = N\psi + pV$$

the Gibbs free enthalpy, where N is the particle number and $V = N/\rho$ the volume. Its density $\gamma = G/N$, which is named *chemical potential*, reads

$$\gamma = \psi + \frac{p}{\rho} = \Psi + \frac{p}{\rho} + \frac{\mu_0\theta}{2\rho}|\nabla\varphi|^2. \tag{26}$$

Henceforth, we restrict our attention to a uniform configuration, i.e. $\nabla\varphi = \mathbf{0}$.

After proving that $\Gamma = \Psi + p/\rho$ depends only on p, θ, φ, we are able to state the Clausius-Clapeyron equation and to achieve the explicit expression of the curve $p = \pi_v(\theta)$. In fact, by virtue of (24) and (26), we have

$$\Gamma(\rho,\theta,\varphi) = c_p\theta(1-\ln\theta) + Lf(\varphi) + k\theta[1-\nu g(\varphi)]\ln\frac{p(\rho,\theta,\varphi)}{p_c} + \frac{p(\rho,\theta,\varphi)}{\rho}.$$

and after substituting (21) into the last term of the right hand side we obtain

$$\Gamma(\rho,\theta,\varphi) = c_p\theta(1-\ln\theta) + Lf(\varphi) + k\theta[1-\nu g(\varphi)]\left(\ln\frac{p(\rho,\theta,\varphi)}{p_c}+1\right) \tag{27}$$

so proving that Γ depends on ρ through p. Hence, we set

$$\tilde{\Gamma}(p,\theta,\varphi) = \Gamma(\rho,\theta,\varphi),$$

and, for later convenience, we introduce the function u given by

$$u(p,\theta) = -\frac{\nu k\theta}{L}\left(\ln\frac{p}{p_c}+1\right). \tag{28}$$

It is worth noting that $u > 0$ whenever $p < p_c/e$ and $\theta < \tau(p_c/e)$. So doing, the Gibbs free enthalpy density can be rewritten in the form

$$\tilde{\Gamma}(p,\theta,\varphi) = c_p\theta(1-\ln\theta) + k\theta\left(\ln\frac{p}{p_c}+1\right) + L[f(\varphi)+u(p,\theta)g(\varphi)].$$

On the vapor pressure curve, i.e. when $p^* = \pi(\theta^*)$, vapor and liquid both coexist as stable phases. As a consequence, $\tilde{\Gamma}$ must take the same value at $\varphi = 0$ and $\varphi = 1$ (see, for instance [7]), namely $\Gamma(p^*,\theta^*,0) = \Gamma(p^*,\theta^*,1)$, which implies

$$f(1) + u(p^*,\theta^*)g(1) = f(0) + u(p^*,\theta^*)g(0).$$

From this condition it follows $u(p^*,\theta^*) = 1$, so that $u(p,\theta)$ looks like the transition parameter according as its value is greater or smaller than 1. In particular, its restriction along the vapor pressure curve can be expressed as a function of either the transition temperature, θ^*, or the transition pressure, p^*, namely

$$u(\pi_v(\theta^*),\theta^*) = -\frac{\nu k\theta^*}{L}\left[\ln\frac{\pi_v(\theta^*)}{p_c}+1\right], \qquad u(p^*,\tau_v(p^*)) = -\frac{\nu k\tau_v(p^*)}{L}\left(\ln\frac{p^*}{p_c}+1\right).$$

Finally, the explicit and customary (see, for instance, [7, 10]) form of the vapor pressure curve (3) is achieved by exploiting the condition $u(p^*,\theta^*) = 1$.

Remark 3. *Assuming* $u(\theta^*,\pi_v(\theta^*)) = 1$ *we find out*

$$p^* = \pi_v(\theta^*) = \frac{p_c}{e}\exp\left(-\frac{L}{k\nu\theta^*}\right). \tag{29}$$

Similarly, by letting $u(\tau(p^*),p^*) = 1$ *we obtain*

$$\theta^* = \tau_v(p^*) = -\frac{L}{k\nu(1+\ln p^* - \ln p_c)}. \tag{30}$$

In the (θ,p)-plane, π_v is a monotone increasing function satisfying the condition

$$\lim_{\theta^*\to+\infty}\pi(\theta^*) = \frac{p_c}{e}.$$

Unfortunately, it is worth noting that the critical temperature θ_c has not a finite value, as a consequence of the assumption ν =const.

References

[1] H.W. Alt and I. Pawlow, A mathematical model of dynamics of non-isothermal phase separation, Physica D, 59 (1992) 389-416.

[2] A. Berti and C. Giorgi, A phase-field model for liquid-vapor transitions, submitted.

[3] M. Brokate and J. Sprekels, Hysteresis and Phase Transitions, Springer, New York, 1996.

[4] G. Caginalp, An analysis of a phase field model of a free boundary, Arch. Rational Mech. Anal., 92 (1986) 205–245.

[5] M. Fabrizio, C. Giorgi and A. Morro, A Thermodynamic approach to non-isotermal phase-field evolution in continuum physics, Physica D, 214 (2006) 144–156.

[6] M. Fabrizio, C. Giorgi and A. Morro, A Continuum theory for first-order phase transitions based on the balance of the structure order, Math. Methods Appl. Sci., 31 (2008) 627–653

[7] E. Fermi, Thermodynamics, Dover, New York, 1956.

[8] E. Fried and M.E. Gurtin, Continuum theory of thermally induced phase transitions based on an order parameter, Physica D 68 (1993) 326–343.

[9] A.E. Green and N. Laws, On a global entropy production inequality, Quart. J. Mech. Appl. Math. 25 (1972), 1–11.

[10] W. Greiner, L. Neise and H. Stöcker, Thermodynamics and statistical mechanics, Springer, Berlin,1994.

[11] A. Karma and W.J. Rappel, Quantitative phase-field modeling of dendritic growth in two and three dimensions, Phys. Rev. E., 57 (1998) 4323–4349.

[12] I. Muller, On the entropy inequality, Arch. Rational Mech. Anal., 26 (1967) 118–141.

[13] O. Penrose and P.C. Fife, Thermodynamically consistent models of phase-field type for the kinetics of phase transitions, Physica D 43 (1990), 44-62.

[14] O. Penrose and P.C. Fife, On the relation between the standard phase-field model and a "thermodynamically consistent" phase-field model, Physica D, 69 (1993) 107-113.

Wave Propagation in Continuously-Layered Media

G. Caviglia[a], *A. Morro*[b]

[a]Mathematics Department,
via Dodecaneso 35, 16146 Genova, Italy.

[b]University of Genoa, DIBE,
via Opera Pia 11a, 16145 Genova, Italy.

1 Introduction

Continuously-layered media are a simple but relevant model of inhomogeneous media and amount to allowing for the dependence of material properties on a single Cartesian coordinate, say z. They are especially suited to describe problems in acoustics, seismic exploration, oceanography, electromagnetism in matter.

In this paper we investigate the propagation of time-harmonic acoustic waves, with angular frequency ω, governed by the differential equation

$$(\Delta + \frac{\omega^2}{c^2(z)})\psi = 0, \tag{1}$$

where c^2 is the derivative of pressure p with respect to mass density ρ, at constant entropy. Quite often (see, e.g., [1]) the differential equation for acoustic waves is taken in the form

$$\rho\nabla \cdot (\frac{1}{\rho}\nabla\psi) + \frac{\omega^2}{c^2(z)}\psi = 0. \tag{2}$$

While (2) reduces to (1) if the density ρ is constant, we believe that (1) is more appropriate to describe acoustic waves even though ρ is not constant [2].

The purpose of this paper is twofold. First to show that appropriate wave propagation problems, associated with (1), can be given the form of a Volterra integral equation. This result is remarkable in that usually integral equations are set up in the Fredholm form. It is an advantage of a Volterra integral equation that the solution to the problem exists and is unique without any boundedness requirements on the parameters involved. Secondly, to determine the reflection and transmission coefficients generated by a layer through functions generated by the Volterra integral equations associated with two particular initial data.

2 Waves in stratified media

Let $\Omega \subseteq \mathbb{R}^3$ be the space region occupied by the body under consideration. A function f, on $\Omega \times \mathbb{R}$, is a plane wave propagating in the direction of the constant vector $\mathbf{q}$, with speed $1/|\mathbf{q}|$, if

$$f(\mathbf{x}, t) = \mathcal{F}(t - \mathbf{q} \cdot \mathbf{x})$$

for some function $\mathcal{F}$, on $\mathbb{R}$, which describe the profile of the wave. Plane wave solutions occur in homogeneous media (see [3, 4]), not in inhomogeneous media. Hence we look for generalized plane waves in the form

$$g(\mathbf{x}, t) = \mathcal{G}(z, t - \mathbf{m} \cdot \mathbf{x}),$$

where $\mathbf{m}$ is perpendicular to the z-axis, for any function $\mathcal{G}$ on $\mathbb{R}^2$. By an appropriate choice of the Cartesian axes we let $\mathbf{m}$ be directed along the x-axis and write

$$g(\mathbf{x}, t) = \mathcal{G}(z, t - \xi x), \qquad \xi \in \mathbb{R}.$$

To simplify the connection with the literature on the subject we let the Fourier transform $\tilde{g}$, with respect to t, be given by

$$\tilde{g}(\mathbf{x}, \omega) = \int_{-\infty}^{\infty} g(\mathbf{x}, t) \exp(i\omega t) dt.$$

Hence we have

$$\tilde{g}(\mathbf{x}, \omega) = \exp(i\omega\xi x)\tilde{\mathcal{G}}(z, \omega).$$

2.1 Acoustic wave

Look at (1) for acoustics. The unknown ψ is the pressure p. Applying the Fourier transform gives

$$\frac{\omega^2}{c^2(z)}\tilde{p} + \Delta\tilde{p} = 0.$$

Substitution of $\tilde{p}(\mathbf{x}, \omega) = \exp(i\omega\xi x)P(z, \omega)$ provides

$$P'' + N(z, \omega)P = 0$$

where a prime stands for ∂_z and

$$N(z, \omega) = \omega^2 (\frac{1}{c^2(z)} - \xi^2).$$

If $N > 0$ we define

$$\kappa = \sqrt{N}.$$

We let N suffer jump discontinuities at $z = 0, L$ and hence take $\Omega_z = (-\infty, 0) \cup (0, L) \cup (L, \infty)$ as the domain of N. Also we assume that N takes constant, positive, possibly-unequal values in the half-spaces $z < 0$, $z > L$. Hence we let

$$\kappa(z) = k_0, \quad z < 0, \qquad \kappa(z) = k_1, \quad z > L.$$

Denote by c_0, c_1 the values of c as $z < 0$ and $z > L$ and let $\beta = c_0\xi$. We have

$$N = k_0^2(1-\alpha), \qquad k_0^2 = \frac{\omega^2}{c_0^2}(1-\beta^2), \qquad \alpha = \frac{c^2 - c_0^2}{c^2(1-\beta^2)}.$$

As a consequence the unknown pressure P on $\Omega_z \times \mathbb{R}$ satisfies the differential equation

$$P'' + k_0^2 P = k_0^2 \alpha(z) P, \qquad z \in \Omega_z. \tag{3}$$

3 Variational formulation

A two-point function G on $(0, L) \times (0, L)$ satisfying the distribution equation

$$\partial_z^2 G(z,s) + k_0^2 G(z,s) = \delta(z-s), \qquad z, s \in (0, L), \tag{4}$$

is a fundamental solution [5] of the operator $P'' + k_0^2 P$. Restrict attention to continuous fundamental solutions, that is

$$G(s_-, s) = G(s_+, s). \tag{5}$$

Since

$$\partial_z^2 G(z,s) + k_0^2 G(z,s) = 0, \qquad z \neq s, \tag{6}$$

integration of (4) with respect to z on $[z_1, z_2] \ni s$, $z_1, z_2 \in (0, L)$, provides

$$\partial_z G(z_2, s) - \partial_z G(z_1, s) + k_0^2 \int_{z_1}^{z_2} G(\tau, s) d\tau = 1.$$

The limit $z_1 \to s_-$ and $z_2 \to s_+$ gives

$$\partial_z G(s_+, s) - \partial_z G(s_-, s) = 1. \tag{7}$$

The distribution equation (4) amounts to

$$\int_0^L G(z,s)(\partial_z^2 + k_0^2)\phi(z) dz = \phi(s), \qquad s \in (0, L), \tag{8}$$

for any test function ϕ on $[0, L]$. The conditions (5)-(7) characterize a set of fundamental solutions in that imply the validity of (8).

Proposition 1. *Any function G, on $(0, L) \times (0, L)$, satisfying (6) and the jump conditions (5), (7) is a fundamental solution of $P'' + k_0^2 P$ on $(0, L)$.*

Proof. For any $s \in (0, L)$, regard the integral on $(0, L)$ as the sum of that on $(0, s)$ and that on (s, L). For each of them we integrate by parts to obtain

$$\int_0^s G(z,s)\phi''(z) dz = G(s_-, s)\phi'(s) - \partial_z G(s_-, s)\phi(s) + \int_0^s \partial_z^2 G(z,s)\phi(z) dz$$

and

$$\int_s^L G(z,s)\phi''(z) dz = -G(s_+, s)\phi'(s) + \partial_z G(s_+, s)\phi(s) + \int_s^L \partial_z^2 G(z,s)\phi(z) dz.$$

Hence, by the continuity (5) of G at $z = s$ we have

$$\int_0^L \partial_z G(z,s)\,\phi'(z)dz = \left[\phi(z)\,\partial_z G(z,s)\right]_{s_-}^{s_+} + \int_0^L \partial_z^2 G(z,s)\,\phi(z)dz.$$

By the continuity of ϕ and the jump condition (7) on $\partial_z G$ we have

$$\left[\phi(z)\,\partial_z G(z,s)\right]_{s_-}^{s_+} = \phi(s).$$

As a consequence, because

$$\partial_z^2 G(z,s) = k_0^2 G(z,s), \quad z \in (0,s) \cup (s,L),$$

we obtain (8)

If, instead, $s \notin [0,L]$ then the analogous procedure involves a zero jump and we have

$$\int_0^L G(z,s)(\partial_z^2 + k_0^2)\phi(z)dz = 0, \qquad s \notin [0,L],$$

for any test function ϕ on $[0,L]$.

Multiply (3) by $G(z,s)$ and (6) by $P(z)$. Take the difference and integrate on $(0,s) \cup (s,L)$. Integration by parts, use of the conditions (5), (7) and some rearrangements give

$$P(z) = k_0^2 \int_0^L [\alpha(s)P(s)G(s,z)]ds - \left[G(s,z)\partial_s P(s) - P(s)\partial_s G(s,z)\right]_{s=0_+}^{s=L_-}. \quad (9)$$

By (6), the solution is a 4-parameter family of fundamental solutions. The continuity of G (see (5)) and the jump condition (7) leave two parameters free. As a consequence we can use the freedom in selecting the two parameters so that the integral equation (9) takes a more convenient form.

3.1 Fundamental solutions

We first look at

$$G_1(z,s) = \begin{cases} \frac{1}{k_0}\sin k_0(s-z), & z \in (0,s), \\ 0, & z \in (s,L). \end{cases} \quad (10)$$

Both the differential equation (6) and the jump conditions (5), (7) hold. Hence G_1 is a fundamental solution.

Upon substitution of (10) for G in (9) we have

$$P(z) = \frac{1}{2}[\frac{P'(0_+)}{ik_0} + P(0_+)]\exp(ik_0 z) + \frac{1}{2}[-\frac{P'(0_+)}{ik_0} + P(0_+)]\exp(-ik_0 z)$$
$$+k_0 \int_0^z \alpha(s)P(s)\sin(k_0(z-s))ds. \quad (11)$$

For any value of k_0, eq. (11) is made operative once $P(0_+)$ and $P'(0_+)$ are fixed.

A function G_2 is now considered in the form

$$G_2(z,s) = \begin{cases} -\dfrac{1}{k_0}\dfrac{\sin(k_0(L-s))}{\sin(k_0 L)}\sin(k_0 z), & z \in (0,s), \\ -\dfrac{1}{k_0}\dfrac{\sin(k_0 s)}{\sin(k_0 L)}\sin(k_0(L-z)), & z \in (s,L). \end{cases} \tag{12}$$

The function G_2 satisfies the differential equation (6) and the jump conditions (5), (7) and hence is a fundamental solution, provided $\sin(k_0 L) \neq 0$. Also, $G_2(z,w)$ vanishes at $z = 0, L$, namely

$$G_2(0_+, s) = G_2(L_-, s) = 0.$$

Moreover, it follows that

$$\partial_z G_2(0_+, s) = \frac{\sin(k_0 s)}{\sin(k_0 L)}, \qquad \partial_z G_2(L_-, s) = -\frac{\sin(k_0(L-z))}{\sin(k_0 L)}.$$

Upon substitution of (12) for G in (9) we find that

$$P(z) = k_0^2 \int_0^L \alpha(\tau)P(\tau)G_2(\tau,z)d\tau + P(L_-)\frac{\sin(k_0 z)}{\sin(k_0 L)} + P(0_+)\frac{\sin(k_0(L-z)}{\sin(k_0 L)}. \tag{13}$$

For any value of k_0, eq. (13) is made operative once $P(0)$ and $P(L)$ are fixed.

A further fundamental solution, say G_3, is applied in [6, 7] and is taken in the form

$$G_3(z,s) = \frac{1}{2ik_0}\exp(ik_0|z-s|).$$

It is immediate to see that (5) holds and that

$$\partial_z G_3(s_+, s) = \frac{1}{2}, \qquad \partial_z G_3(s_-, s) = -\frac{1}{2}$$

so that (7) holds too. The corresponding integral equation for P follows by replacing G_3 for G in (9).

A variant of G_3 is considered in [8]. In our notation the function is taken as

$$G_4(z,s) = \frac{1}{2i|k_0|}\exp(i|k_0||z-s|).$$

Again it follows that the differential equation (6) and the conditions (5), (7) hold. Hence G_4 is a fundamental solution too.

The analogue of G_1 is now considered in the form

$$G_5(s,z) = \begin{cases} 0, & s \in (0,z) \\ \frac{1}{k_0}\sin k_0(s-z), & s \in (z,L) \end{cases} \tag{14}$$

Both the differential equation (6) and the jump conditions (5), (7) hold. Hence also G_5 is a fundamental solution. Since G_5 vanishes on $(0,z)$, (9) provides

$$P(z) = k_0^2 \int_z^L [\alpha(s)P(s)G_5(s,z)]ds - G_5(L,z)P'(L_-) + P(L_-)\partial_s G_5(L,z).$$

Now,

$$-G_5(L,z)P'(L_-) + P(L_-)\partial_s G_5(L,z) = \frac{1}{2}[P(L_-) - \frac{P'(L_-)}{ik_0}]\exp(ik_0(L-z)) + \frac{1}{2}[P(L_-) + \frac{P'(L_-)}{ik_0}]\exp(-ik_0(L-z)).$$

Hence we have

$$P(z) = k_0^2 \int_z^L [\alpha(s)P(s)G_5(s,z)]dy = \frac{1}{2}[P(L_-) - \frac{P'(L_-)}{ik_0}]\exp(ik_0(L-z)) + \frac{1}{2}[P(L_-) + \frac{P'(L_-)}{ik_0}]\exp(-ik_0(L-z)). \quad (15)$$

For any value of k_0, eq. (15) is made operative once $P(L_-)$ and $P'(L_-)$ are chosen.

Equation (13), and the analogous one for G_3 and G_4, are Fredholm equations for P. Equations (11) and (15), instead, are Volterra's integral equations of the second kind. They involve, as parameters, two values among $P(0_+), P(L_-), P'(0_+), P'(L_-)$. In a RT problem such values are unknown. Hence, as we show in a moment, the application of the integral equations to RT problems requires an appropriate setting. Henceforth we apply (11).

3.2 Existence and uniqueness

According as $P'(0_+) = 0$ or $P(0_+) = 0$ eq. (11) takes the form

$$P(z) = \phi(z) + \int_0^z A(z,s)P(s)ds, \qquad z \in [0,L], \quad (16)$$

where

$$\phi(z) = P(0_+)\cos k_0 z \quad \text{or} \quad \phi(z) = \frac{P'(0_+)}{k_0}\sin k_0 z, \qquad A(z,s) = k_0\,\alpha(s)\sin k_0(z-s).$$

By [9], *if ϕ is a continuous function on $[0,L]$ and $A(z,s)$ is a continuous function of (z,s), on $S = \{(z,s)|\ 0 \le s \le z \le L\}$, then there is a unique solution of the Volterra integral equation (16).* Here A and ϕ are continuous and hence existence and uniqueness of P follows. Indeed, by the linearity of (11), the solution to any pair of initial values $P(0_+), P'(0_+)$ is the sum of the solutions associated with $P'(0_+) = 0$ and $P(0_+) = 0$.

It is remarkable that existence and uniqueness of P hold without any requirement on the parameters and functions involved. We might have used a Fredholm integral equation as it happens if G is taken in the form (12). In such a case existence and uniqueness are shown to hold [9] provided there is a positive continuous scalar function θ on $[0,L]$ such that

$$\theta(z)\int_0^L \frac{|A(z,s)|}{\theta(s)}ds < 1, \qquad z \in [0,L]. \quad (17)$$

The inequality (17) can be viewed as a smallness condition on A which guarantees existence and uniqueness. In [7] the Fredholm equation is considered with

$$A(z,s) = \frac{k_0}{2i}\exp(ik_0|z-s|)\alpha(s).$$

The condition (17) for existence and uniqueness then holds if

$$|k_0\alpha|L \leq 2.$$

4 A basis of solutions

Let f_1, f_2 be solutions to (3) on $[0, L]$ with initial values

$$f_1(0) = 1, \quad f_1'(0) = 0, \qquad f_2(0) = 0, \quad f_2'(0) = 1.$$

They are a basis (or a fundamental system) of solutions in that they are linearly independent. For, the Wronskian $W(z) = f_1(z)f_2'(z) - f_2(z)f_1'(z)$ is constant and hence is constantly equal to the initial value, $W(0) = 1$. The general solution P to (3) is then a linear combination of f_1, f_2, that is

$$P(z) = Af_1(z) + Bf_2(z), \qquad z \in (0, L),$$

where $A, B \in \mathbb{C}$. Both f_1 and f_2 may be determined by means of the appropriate form of the integral equation (11), namely

$$f_1(z) = \cos k_0 z + k_0 \int_0^z \alpha(s) f_1(s) \sin k_0(z - s)\, ds, \tag{18}$$

$$f_2(z) = \frac{1}{k_0} \sin k_0 z + k_0 \int_0^z \alpha(s) f_2(s) \sin k_0(z - s)\, ds. \tag{19}$$

4.1 Successive approximations

The successive-approximations method is the way for proving existence and uniqueness of the solution to (16) [9]. For practical purposes, we now examine some properties of the successive approximations to f.

Let

$$f^{(1)}(z) = \phi(z), \qquad f^{(j+1)}(z) = \phi(z) + \int_0^z A(z, s) f^{(j)}(s) ds, \quad j \in \mathbb{N}. \tag{20}$$

Also, let

$$|\phi| \leq m_1, \qquad |A(z, s)| \leq |k_0\alpha(s)| \leq m_2.$$

Hence

$$|f^{(2)}(z) - f^{(1)}(z)| = \left| \int_0^z A(z, z_1) f^{(1)}(z_1) dz_1 \right| \leq m_1 m_2 z.$$

For any value of $j \in \mathbb{N}$,

$$f^{(j+1)}(z) - f^{(j)}(z) = \int_0^z \int_0^{z_j} \cdots \int_0^{z_3} A(z, z_j) \cdots A(z, z_2)[f^{(2)}(z_2) - f^{(1)}(z_2)] dz_2 \cdots dz_j$$

whence the estimate of $|f^{(j+1)}(z) - f^{(j)}(z)|$, use of $|f^{(2)}(z_2) - f^{(1)}(z_2)| \leq m_1 m_2 z_2$ and the $j - 1$ integrations give

$$|f^{(j+1)}(z) - f^{(j)}(z)| \leq m_1 \frac{(m_2 z)^j}{j!}, \quad j \in \mathbb{N}. \tag{21}$$

The identity

$$f^{(j+p)}(z) - f^{(j)}(z) = \sum_{h=j}^{j+p-1} f^{(h+1)}(z) - f^{(h)}(z)$$

and use of (21) give

$$|f^{(j+p)}(z) - f^{(j)}(z)| \leq \sum_{h=j}^{\infty} |f^{(h+1)}(z) - f^{(h)}(z)| \leq m_1 \sum_{h=j}^{\infty} \frac{(m_2 L)^h}{h!}$$

whence

$$|f^{(j+p)}(z) - f^{(j)}(z)| \leq m_1 \frac{(m_2 L)^j}{j!} \exp(m_2 L).$$

As a consequence, the limit of the sequence $f^{(j)}$ exists and, by (20), is the solution f, at any point $z \in [0, L]$. Taking the limit as $p \to \infty$ we have the estimate

$$|f(z) - f^{(j)}(z)| \leq m_1 \frac{(m_2 L)^j}{j!} \exp(m_2 L) \tag{22}$$

for the error $|f(z) - f^{(j)}(z)|$. The estimate shows how the error increases with the wave-number k_0 and the maximum value of $|\alpha|$ ($m_2 \geq |k_0 \alpha|$), and the thickness L of the layer.

5 Reflection-transmission problem

Equation (11) involves the initial values $P(0_+), P'(0_+)$. In an RT problem, instead, neither $P(0_+)$ nor $P'(0_+)$ are known. Hence, to apply (11) to an RT problem we need a formulation where at least two parameters occur which can be associated with $P(0_+)$ and $P'(0_+)$. The reflection and transmission coefficients R, T are among such parameters. Once the appropriate jump conditions are applied, the pertinent parameters, and hence also $P(0_+)$ and $P'(0_+)$, are determined.

Since $N = k_0^2$ as $z < 0$ and $N = k_1^2$ as $z > L$, the solution P to (3), is a linear combination of $\exp(ik_0 z), \exp(-ik_0 z)$, as $z < 0$, whereas it is a linear combination of $\exp(ik_1(z-L)), \exp(-ik_1(z-L))$, as $z > L$. Denote by c_0 and c_1 the wave speed c as $z < 0$ and $z > L$. Likewise, let θ_0 and θ_1 be the angle between the direction of the plane wave and the z-axis as $z < 0$ and $z > L$. Hence by Snell's law,

$$\frac{1}{c_0} \sin\theta_0 = \frac{1}{c_1} \sin\theta_1.$$

Moreover, k_0 and k_1 are given by

$$k_0 = \frac{\omega}{c_0} \cos\theta_0, \qquad k_1 = \frac{\omega}{c_1} \cos\theta_1.$$

Let us consider the RT problem associated with the incident wave

$$P_{\text{inc}}(z) = \exp(ik_0 z), \qquad z < 0$$

Hence we look for the solution P in the form

$$P(z) = \begin{cases} \exp(ik_0 z) + R\exp(-ik_0 z), & z < 0, \\ Af_1(z) + Bf_2(z), & z \in (0, L), \\ T\exp(ik_1(z - L)), & z > L, \end{cases}$$

where $R, A, B, T \in \mathbb{C}$. Of course,

$$P'(z) = \begin{cases} ik_0[\exp(ik_0 z) - R\exp(-ik_0 z)], & z < 0, \\ Af_1'(z) + Bf_2'(z), & z \in (0, L), \\ ik_1 T\exp(ik_1(z - L)), & z > L. \end{cases}$$

It is natural to regard R as the reflection coefficient and T as the transmission coefficient, relative to the wave in $z > L$. The coefficients A, B characterize the solution in the layer. If the transmitted wave occurs in $z > 0$, which means that the thickness of the layer is zero, then the same calculation applies provided $T\exp(-ik_1 L)$ is viewed as the transmission coefficient.

We let ρ suffer jump discontinuities at $z = 0, L$ and take the standard jump conditions

$$[\![P]\!] = 0, \qquad [\![\frac{1}{\rho}P']\!] = 0, \qquad z = 0, L, \tag{23}$$

where $[\![\quad]\!]$ denotes the jump at the pertinent value of z,

$$[\![P]\!](z) = P(z_+) - P(z_-).$$

Also let

$$J_0 = \frac{\rho(0_+)}{\rho(0_-)}, \qquad J_1 = \frac{\rho(L_-)}{\rho(L_+)}.$$

The continuity of P'/ρ gives

$$J_0 P'(0_-) = P'(0_+), \qquad P'(L_-) = J_1 P'(L_+).$$

By applying (23) we obtain

$$1 + R = A, \qquad Af_1(L) + Bf_2(L) = T,$$

$$ik_0 J_0(1 - R) = B, \qquad Af_1'(L) + Bf_2'(L) = ik_1 J_1 T.$$

Letting

$$F_m = f_m(L) + i\frac{f_m'(L)}{k_1 J_1}, \qquad m = 1, 2,$$

we find that

$$R = -\frac{F_1 + ik_0 J_0 F_2}{F_1 - ik_0 J_0 F_2}, \qquad T = 2ik_0 J_0 \frac{F_1 f_2 - F_2 f_1}{F_1 - ik_0 J_0 F_2},$$

$$A = -\frac{2ik_0 J_0 F_2}{F_1 - ik_0 J_0 F_2}, \qquad B = \frac{2ik_0 J_0 F_1}{F_1 - ik_0 J_0 F_2}.$$

6 Conclusions

Various fundamental solutions for the differential equation (3) are established. While G_1 and G_5 provide Volterra integral equations, G_2, G_3, G_4 enter Fredholm integral equations. The Volterra integral equations look unusual in the literature. They have the advantage that the solution exists and is unique without any boundedness requirements on the parameters involved. The base functions f_1, f_2 are associated with two particular initial-value problems, eqs. (18), (19). A classical RT problem, associated with a continuously-layered layer, is then solved in terms of the base functions f_1, f_2.

Acknowledgement

The research leading to this paper has been supported by the Italian MIUR through the Research Project COFIN 2005 *Mathematical Models and Methods in Continuum Physics.*

References

[1] P.A. Martin, *Acoustic scattering by inhomogeneous spheres,* J. Acoust. Soc. Am. 111, 2013-2018 (2002).

[2] G. Caviglia, A. Morro, *Sound-wave equation in the acoustic approximation,* Ist. Lombardo Sci. Lett., to appear.

[3] G. Caviglia, A. Morro, B. Straughan, *Reflection and refraction at a fluid-porous medium interface,* J. Acoust. Soc. Am. 92, 1113-1119 (1992).

[4] G. Caviglia, A. Morro, B. Straughan, *Thermoelasticity at cryogenic temperatures,* Int. J. Non-Linear Mech. 27, 251-263 (1992).

[5] W. Cheney, *Analysis for Applied Mathematics,* Springer, 2000, p. 273.

[6] K. A. Innanen and A. B. Weglein, *Construction of absorptive/dispersive wave fields with the forward scattering series,* J. Seismic Exploration 12, 53-74 (2003).

[7] A. B. Weglein, F. V. Araujo, P. M. Carvalho, R. H. Stolt, K. H. Matson, R. T. Coates, D. Corrigan, D. J. Foster, S. A. Shaw, H. Zhang, *Inverse scattering series and seismic exploration,* Inverse Problems 19, R27-R83 (2003).

[8] M. Razavy, *Determination of the wave velocity in an inhomogeneous medium from the reflection coefficient,* J. Acoust. Soc. Am. 58, 956-963 (1975).

[9] W.T. Reid, *Ordinary Differential Equations,* Wiley, 1971, p. 57.

Nonlinear Stability for Reaction-Diffusion Models

*G. Mulone**

Dipartimento di Matematica e Informatica
Città Universitaria,
viale A. Doria 6, 95125, Catania, Italy.

Dedicated to Brian Straughan on the occasion of his 60th birthday.

Abstract

Some linear and nonlinear stability results for constant solutions to reaction-diffusion models are presented. Optimal stability results with the reduction method (based on the classical theory of eigenvalues-eigenvectors) are recalled. Global stability has been obtained for the positive equilibria of two competition models and for the endemic state of an epidemic model with diffusion.

1 Introduction

Nonlinear reaction-diffusion systems are well suited to model a wide range of physical, chemical, biological pattern formation processes, and biomedical applications. The study of the stability-instability of a given basic solution is very important to understand the real world (see, for instance, Murray 2002, 2003, Anderson et al. 1981, Straughan 2004, Quinlan and Straughan 2005, Webber and Straughan 2006).

Let us consider a reaction-diffusion system

$$u_t = D\Delta u + Lu + N(u_1, \ldots, u_n) \tag{1}$$

$u(x,t) = (u_1, u_2, \ldots, u_n)^T$, with $x \in \Omega$, a bounded domain of $\mathbb{R}^m$, with suitable boundary conditions (usually Dirichlet or Neumann boundary conditions), $t \in (0, +\infty)$. Assume that u is an element of a Hilbert space $\mathcal{H} = L^2(\Omega)$. $D = D_{ij}$ and $L = L_{ij}$, $i, j = 1, 2, \ldots, n$, are constant matrices (depending on some *physical or biological* parameters), Δ is the m-dimensional Laplacian and $N = (N_1, \ldots, N_n)^T$ represents the nonlinearities (in some problems N_i, $i = 1, 2, \ldots, n$, are polynomial in U_j and $D_{ij} = \delta_{ij} D_j$, with positive D_j). Moreover, we suppose that $\tilde{U}$ is a constant solution of the reaction - diffusion system with appropriate boundary conditions.

*Email: mulone@dmi.unict.it

Let us consider the perturbation equations to $\tilde{U}$:

$$U_t = D\Delta U + L'U + N'(U_1, \dots, U_n) \tag{2}$$

with initial condition

$$U(0) = U_0 \tag{3}$$

and zero boundary conditions (Dirichlet or Neumann). In (2) L' and N' are the linear and nonlinear operators. The perturbation $U = (U_1, U_2, \dots, U_n)^T$, and the initial condition U_0 are elements of $\mathcal{H}$.

Here we suppose that the initial value problem (2)–(3), with suitable boundary conditions, is well posed in $\mathcal{H}$ and the solution exists globally at least for small initial data.

In order to study the stability-instability problem of $\tilde{U}$ the main classical methods are the *linearised stability-instability method* and the Lyapunov method. The first method is based on the study of the spectrum of the linear operator of the system. It provides a *critical parameter* R_c, for a given parameter R, above (or below) which $\tilde{U}$ is unstable. The *Lyapunov method* (with a Lyapunov function E) provides a *critical nonlinear Lyapunov parameter* R_E below (or above) which the basic solution $\tilde{U}$ is nonlinearly stable, (see Henry 1981, Smoller 1994, Flavin and Rionero 1996, Straughan 2004).

In general, we have $R_E \leq R_c$. If, in particular, $R_E = R_c$, one obtains *necessary and sufficient* conditions of linear and nonlinear (conditional or global) stability and we say that the Lyapunov function E is *optimal.* This, for instance, happens if the linear operator is symmetric or symmetrizable in the scalar product of the Hilbert space (see Galdi and Straughan 1985).

A fair amount of attention has been given to the application of Lyapunov methods to reaction diffusion systems (Henry 1981). A nonlinear stability analysis for reaction - diffusion systems has also been given in Fitzgibbon et al. 1997, Rionero 2006, by introducing particular Lyapunov functions.

For ODE systems a linearization principle holds, the Hartman - Grobman theorem, while for PDE system it has to prove case by case. It also holds for reaction diffusion systems (see, for example, Henry 1981, pp. 98–100, Smoller 1994). It gives stability conditions up to the linear criticality but it does not give a computable radius of attractivity for initial data in terms of the physical parameters of the problem. In many physical systems, in PDEs case, the *classical energy*, $E_0(t) = \frac{1}{2}\|U\|^2$, is used as a good Lyapunov function to control the stability. In these cases one generally obtains $R_{E_0} \leq R_c$, in particular, if the linear operator A is *symmetric* (Straughan 2004), one has $R_{E_0} = R_c$. If A has is not symmetric, then, usually, $R_{E_0} < R_c$ and new Lyapunov functions, different from the classical energy E_0, must be introduced to improve the nonlinear critical stability thresholds (see Galdi and Straughan 1985, Mulone and Rionero 1997, 2003, Payne and Straughan 1999a, 1999b, Straughan 2001, 2002).

Now the following questions arise:

(1) Is there any procedure to build an *optimal* Lyapunov function to obtain nonlinear (at least conditional) stability?

(2) Are there Lyapunov functions which give *global* stability results?

In the next sections, we study these questions for some reaction-diffusion models.

2 Linear and nonlinear stability by the reduction method

In this section we answer question (1). For this, firstly let us recall the results in the ODEs case.

Assume that the perturbation system to an equilibrium is given by

$$U' = AU + N(U) \tag{4}$$

where A is a constant matrix, $U \in \mathbb{R}^n$ and $N(U)$, sufficiently smooth, is the nonlinear term.
The canonical reduction method based on the classical eigenvalues-eigenvectors problem can be applied. One introduces a change of dependent variables by means of a nonsingular transformation matrix Q, $U = QV$, $V = Q^{-1}U$, to obtain a new (topologically equivalent) system

$$\dot{V} = BV + \overline{N}(V),$$

where $B = Q^{-1}AQ$ is a *similar matrix* to A (it is in a diagonal or a general Jordan form), $\overline{N}(V) = Q^{-1}N(QV)$ (see, e.g., Bellman 1953, pp. 80–82, Hale 1969, Hartman 1982, Beltrami 1997). We thus define the *optimal* Lyapunov function

$$E := \frac{1}{2}[V_1^2 + V_2^2 + \cdots + V_n^2].$$

In the case of reaction-diffusion system (2), we generalize the previous approach by introducing a change of dependent variables, connected with the *first eigenvalue* of the linear operator. Thus, we build as *optimal Lyapunov* function (at least for the linear problem) *the classical energy of the new canonical fields* (Mulone and Straughan 2006, Lombardo et al. 2008):

$$E(t) = \frac{1}{2}\|V\|^2,$$

where $\|\cdot\|$ is the norm in the Hilbert space $\mathcal{H}$. For particular nonlinearities and dimensions of the space domain, sometimes we have to add to $\frac{1}{2}\|V\|^2$ a complementary term. Now we define

$$E(t) = \frac{1}{2}\|V\|^2 + \tilde{b}E_2(t),$$

where $\tilde{b} \geq 0$ and $E_2(t)$ controls the nonlinearities (see Straughan 2004).

The problem of global nonlinear stability is more difficult and it is an open problem for a general system. However, for particular nonlinearities, it is possible to define *suitable* Lyapunov functions which give global stability.

3 Applications and global stability results

Here we apply the reduction method to three reaction-diffusion systems: two ecological problems and an epidemic problem with diffusion. For these systems we obtain also global stability results.

3.1 A 2-dimensional competition model

Let us consider a 2-competing system with diffusion (symmetric case):

$$u_{1,t} = \mu\Delta u_1 + u_1(1 - u_1 - \alpha u_2), \quad u_{2,t} = \mu\Delta u_2 + u_2(1 - \alpha u_1 - u_2), \tag{5}$$

where the spatial domain is a bounded domain $\Omega \subset \mathbb{R}^m$, and $t \in (0, +\infty)$. α is a positive competition coefficient and the competing species satisfy the inequality $u_i \geq 0$. (For an existence theorem and the *positivity and boundeness* of the solution, see Smoller 1994, Chap. 14, Σ F).
The system admits four constant equilibria: $(0,0), (0,1), (1,0), (\overline{u}, \overline{u})$, where $\overline{u} = (1+\alpha)^{-1}$.
Here we study the stability of the positive coexistence solution $(\overline{u}, \overline{u})$. The associated ODE system is

$$\dot{u}_1 = u_1(1 - u_1 - \alpha u_2), \quad \dot{u}_2 = u_2(1 - \alpha u_1 - u_2).$$

The eigenvalues of the Jacobian matrix evaluated at $(\overline{u}, \overline{u})$ are $\lambda_1 = -1$, $\lambda_2 = (\alpha - 1)\overline{u}$. Thus

$$\alpha > 1$$

is sufficient for instability. The perturbations system to $(\overline{u}, \overline{u})$ can be written as

$$\dot{U}_1 = -(\overline{u} + U_1)(U_1 + \alpha U_2), \quad \dot{U}_2 = -(\overline{u} + U_2)(\alpha U_1 + U_2).$$

By using the Lyapunov function

$$E = \sum_{i=1}^{2} (U_i - \overline{u} \ln(1 + \frac{U_i}{\overline{u}}))$$

we have: $\dot{E} = -U_1^2 - 2\alpha U_1 U_2 - U_2^2$ and $\alpha < 1$ implies global nonlinear asymptotical stability in $\mathbb{R}^2_+$ (i.e., $U_i + \overline{u} > 0$).
The perturbation equations of the original (PDE) system are:

$$U_{1,t} = \mu\Delta U_1 - (\overline{u} + U_1)(U_1 + \alpha U_2), \quad U_{2,t} = \mu\Delta U_2 - (\overline{u} + U_2)(\alpha U_1 + U_2). \tag{6}$$

We denote by $-\xi$ the first eigenvalue of the Laplacian ($\xi = 0$ for Neumann boundary conditions, $\xi > 0$ for Dirichlet b.c. Moreover, $\xi > 0$ in the case of mixed (Robin) boundary conditions), we have instability if

$$(\alpha - 1)\overline{u} - \mu\xi > 0.$$

In the case of zero Neumann boundary conditions, we note that $\xi > 0$, if we are interested in linear instability of the positive equilibrium that is *solely spatially dependent* (Murray 2003, pp. 82–84).

Since the linear operator of the system (6) is symmetric in the usual scalar product of $L^2(\Omega)$, by using the energy:

$$E_1 = \frac{1}{2}(\|U_1\|^2 + \|U_2\|^2),$$

we have:

$$\dot{E}_1 \le -\mu(\|\nabla U_1\|^2 + \|\nabla U_2\|^2) - 2\overline{u}E_1 - 2\alpha\overline{u}(U_1, U_2) - \int_\Omega (U_1^3 + \alpha U_1^2 U_2 + \alpha U_2^2 U_1 + U_2^3) d\Omega.$$

It can be shown that

$$(\alpha - 1)\overline{u} - \mu\xi < 0\,,$$

implies nonlinear (conditional) stability.
Following Hastings 1978, we use the new Lyapunov function

$$E = \int_\Omega \sum_{i=1}^{2} \left(U_i - \overline{u}\ln(1 + \frac{U_i}{\overline{u}}) \right) d\Omega\,,$$

it is easy to check that

$$\dot{E} = -\int_\Omega (U_1^2 + 2\alpha U_1 U_2 + U_2^2)\, d\Omega - \mu\overline{u} \int_\Omega [\frac{\nabla U_1 \cdot \nabla U_1}{(\overline{u} + U_1)^2} + \frac{\nabla U_2 \cdot \nabla U_2}{(\overline{u} + U_2)^2}]\, d\Omega\,.$$

¿From last identity, in the case of Neumann b.c., if $\alpha < 1$ (same condition of linearised stability), we have *global* nonlinear asymptotical stability in $\mathbb{R}^2_+$ (i.e., $U_i + \overline{u} > 0$). In the case of Dirichlet or mixed boundary conditions, we also have global stability in the case $\alpha \le 1$, while, for $\alpha > 1$, the problem is more difficult and, up to now, the best result is the conditional nonlinear stability obtained above.

We note that a similar result can be obtained in the case of two positive competition coefficients α and β, with $\alpha < 1$ and $\beta < 1$, $\alpha \ne \beta$.

3.2 The asymmetric May-Leonard model with diffusion

The asymmetric May-Leonard system for three competing species with diffusion is given by:

$$\begin{aligned} u_{1,t} &= \mu\Delta u_1 + u_1(1 - u_1 - \alpha_1 u_2 - \beta_1 u_3) \\ u_{2,t} &= \mu\Delta u_2 + u_2(1 - \beta_2 u_1 - u_2 - \alpha_2 u_3) \\ u_{3,t} &= \mu\Delta u_3 + u_3(1 - \alpha_3 u_1 - \beta_3 u_2 - u_3) \end{aligned} \tag{7}$$

for $\mathbf{x} \in \Omega$, $t \in (0, \infty)$ (see Chi et. al. 1998, May and Leonard 1975, Cantrell and Cosner 2003). We suppose we have assigned boundary conditions (of Dirichlet or Neumann type) on $\partial\Omega \times (0, \infty)$, where Ω is a bounded domain of $\mathbb{R}^3$ (or $\mathbb{R}^2$), u_i denotes the nonnegative species density of the ith competitor. The positive constants μ, α_i and β_i, ($i = 1, 2, 3$), are the diffusion rate and the competition coefficients, respectively, and $\alpha_i < 1 < \beta_i$. Chi et. al. 1998, showed that, in absence of diffusion, the system (7) has a unique positive equilibrium $P = (p_1, p_2, p_3) = (\frac{H_1}{H}, \frac{H_2}{H}, \frac{H_3}{H})$, where

$$\begin{aligned} &H_1 = A_1A_2 + A_2B_3 + B_3B_1, \quad H_2 = A_2A_3 + A_3B_1 + B_1B_2, \\ &H_3 = A_3A_1 + A_1B_2 + B_2B_3, \\ &H = B_1B_2B_3 + B_1B_2 + B_2B_3 + B_3B_1 + A_1B_2 + A_2B_3 \\ &\quad + A_3B_1 + A_1A_2 + A_2A_3 + A_3A_1(1 - A_2) \end{aligned} \tag{8}$$

and $A_i = 1 - \alpha_i > 0$, $B_i = \beta_i - 1 > 0$. P is globally asymptotically stable provided $A_1A_2A_3 > B_1B_2B_3$. In particular, in the *symmetric* case, i.e., $\alpha_i = \alpha$, $\beta_i = \beta$, May and Leonard 1975, it is easy to check that the positive constant equilibrium of system (7) is given by $P = (\frac{1}{h}, \frac{1}{h}, \frac{1}{h})$, with $h = 1+\alpha+\beta$, it is globally stable in $\mathbb{R}^3_+$ if $\alpha+\beta < 2$. The positive solution P is also a solution of the system (7). It is obtained with zero Neumann conditions or assigned Dirichlet conditions. Our goal is to study the stability of P. To this end, we write the perturbation equations to P where $u_i = p_i + U_i$, $i = 1, 2, 3$,

$$\begin{aligned} U_{1,t} &= \mu\Delta U_1 - (p_1 + U_1)(U_1 + \alpha_1 U_2 + \beta_1 U_3) \\ U_{2,t} &= \mu\Delta U_2 - (p_2 + U_2)(\beta_2 U_1 + U_2 + \alpha_2 U_3) \\ U_{3,t} &= \mu\Delta U_3 - (p_3 + U_3)(\alpha_3 U_1 + \beta_3 U_2 + U_3)\,. \end{aligned} \tag{9}$$

It is easy to check that the linear stability condition is given by

$$\nu := \frac{1}{2}(1 - p_1 - p_2 - p_3) - \mu\xi = \frac{1}{2H}(B_1B_2B_3 - A_1A_2A_3) - \mu\xi < 0, \tag{10}$$

where ξ is the first eigenvalue of the the operator $-\Delta$ with the appropriate boundary conditions ($\xi = 0$ for Neumann b.c., $\xi > 0$ for Dirichlet or mixed b.c.).

Concerning nonlinear stability, we note that for some α_i and β_i, by using the classical energy

$$E_0 = \frac{1}{2}[\|U_1\|^2 + \|U_2\|^2 + \|U_3\|^2],$$

we can have a conditional nonlinear stability threshold which is *different* from the linear condition (10) (this happens for instance when $\alpha_1 = 0.2, \alpha_2 = \alpha_3 = 0.3, \beta_1 = 1.1, \beta_2 = 1.2, \beta_3 = 1.5$). In order to achieve the same stability threshold as that for the linear case, we use the reduction method. To this end, we define the matrix $A_\xi = -\mu\xi I + J(P)$, where I is the 3×3 identity matrix and $J(P)$ is the Jacobian matrix at P

$$J(P) = \begin{pmatrix} -p_1 & -\alpha_1 p_1 & -\beta_1 p_1 \\ -\beta_2 p_2 & -p_2 & -\alpha_2 p_2 \\ -\alpha_3 p_3 & -\beta_3 p_3 & -p_3 \end{pmatrix}.$$

The eigenvalues λ_i $(i = 1, 2, 3)$ of the matrix A_ξ are $\lambda_1 = -\mu\xi - 1$, $\lambda_{2,3} = \nu \pm i\omega$, where ν is given by (10) and $\omega := \frac{1}{2}\sqrt{4p_1p_2p_3H - (p_1 + p_2 + p_3 - 1)^2}$.

By computing a transformation matrix Q and introducing the new fields $V = Q^{-1}U$, (see Mulone et. al. 2008) we find the new (topologically equivalent to (9)) system:

$$\begin{aligned} V_{1,t} &= \mu\Delta V_1 - V_1 + \overline{N}_1 \\ V_{2,t} &= \mu\Delta V_2 + (\nu + \mu\xi)V_2 + \omega V_3 + \overline{N}_2 \\ V_{3,t} &= \mu\Delta V_3 - \omega V_2 + (\nu + \mu\xi)V_3 + \overline{N}_3, \end{aligned} \tag{11}$$

where $\overline{N}_1, \overline{N}_2, \overline{N}_3$ are the new nonlinear terms.

Now we define the energy

$$E_1 = \frac{1}{2}[\|V_1\|^2 + \|V_2\|^2 + \|V_3\|^2]\,.$$

The evolution equation for E_1 is found to be

$$\begin{aligned}\dot{E}_1 &= -\mu(\|\nabla V_1\|^2 + \|\nabla V_2\|^2 + \|\nabla V_3\|^2) - \|V_1\|^2 \\ &+(\nu + \mu\xi)(\|V_2\|^2 + \|V_3\|^2) + (\overline{N}_1, V_1) + (\overline{N}_2, V_2) + (\overline{N}_3, V_3).\end{aligned}$$

¿From this we easily obtain the nonlinear stability condition $\nu < 0$. Thus, our nonlinear stability threshold coincides with that found from the linearised theory. The nonlinear terms can be controlled as in Mulone and Straughan 2008. It is found that (when a Poincaré inequality holds) there exists a positive constant η_0, depending on α, β, μ, ξ and a positive Sobolev constant c_0, such that $E_1(t)$ decays exponentially fast provided $E_1(0) < \eta_0$.

As concerns the global stability (see Hastings 1978 and the result obtained by Pao 2004 for Neumann boundary conditions, in a more general context), as before, we define the new Lyapunov function

$$E = \int_\Omega \sum_{i=1}^{3} \left(U_i - p_i \ln(1 + \frac{U_i}{p_i}) \right) d\Omega \,.$$

It is easy to check that

$$\dot{E} = - \int_\Omega [U_1^2 + U_2^2 + U_3^2 + (\alpha_1 + \beta_2)U_1U_2 + (\alpha_2 + \beta_3)U_1U_3 + (\alpha_3 + \beta_1)U_2U_3)]\, d\Omega$$
$$-\mu \int_\Omega \sum_{i=1}^{3} p_i \frac{\nabla U_i \cdot \nabla U_i}{(p_i + U_i)^2} d\Omega \,.$$

¿From last identity, if $\alpha_1 + \beta_2 < 2$, and

$$(\alpha_1 + \beta_2)^2 + (\alpha_2 + \beta_3)^2 + (\alpha_3 + \beta_1)^2 - (\alpha_1 + \beta_2)(\alpha_2 + \beta_3)(\alpha_3 + \beta_1) - 4 < 0\,,$$

we have global stability. In particular, in the symmetric May-Leonard case ($\alpha_i = \alpha$, $\beta_i = \beta$), in the case of Neumann boundary conditions, we obtain global nonlinear asymptotical stability in $\mathbb{R}^3_+$ (i.e., $p_i + U_i > 0$) if $\alpha + \beta < 2$. In the case of Dirichlet b. c., if $\alpha + \beta \leq 2$ we obtain global stability, if $\alpha + \beta > 2$ we have a similar result as before: the best (conditional) nonlinear stability up to the linear criticality is that given by the classical energy: $\dfrac{\alpha + \beta - 2}{2h} - \mu\xi < 0\,.$

3.3 An epidemic model with diffusion

Mulone et al. 2007 have considered the epidemic model with diffusion and cross-diffusion

$$S_t = a\Delta S + c\Delta I + \mu - \mu S - \beta SI, \quad I_t = c\Delta S + a\Delta I - (\mu + \epsilon)I + \beta SI. \qquad (12)$$

in $\Omega \times (0, \infty)$, where $S(t, \mathbf{x})$ and $I(t, \mathbf{x})$ are the densities of susceptible individuals and infectious individuals of a biological population at the spatial position $\mathbf{x}$ and the time t, respectively, μ is the recruitment rate of the population and the per capita death rate of the population ($1/\mu$ is the mean lifetime), β is the disease transmission coefficient (contact rate) and ϵ is the disease-induced death rate ($1/\epsilon$ is the average infectious period), a is a diffusion term and c is a cross-diffusion term, $0 \leq c < a$.

Rescaling the variables by introducing

$$t^* = t(\mu+\epsilon), \quad \mathbf{x}^* = \sqrt{\frac{\mu+\epsilon}{a}}\mathbf{x}, \quad R_0 = \frac{\beta}{\mu+\epsilon}, \quad p = \frac{c}{a}, \quad \lambda = \frac{\mu}{\mu+\epsilon},$$

we obtain (omitting the asterisks)

$$S_t = \Delta S + p\Delta I + \lambda(1-S) - R_0 SI, \quad I_t = p\Delta S + \Delta I - I + R_0 SI. \tag{13}$$

We note that $0 \leq p < 1$ and $0 < \lambda < 1$. The basic reproduction number R_0 is used as the threshold quantity that determines *whether a disease can invade* a population.

The equilibria are the disease-free $H_0 \doteq (1,0)$, for any R_0, and, for $R_0 > 1$, the endemic

$$H_1 = (S_e, I_e) = (\frac{1}{R_0}, \lambda(1 - \frac{1}{R_0})).$$

Mulone et al. 2007 have proved that, by choosing the *classical energy* of perturbation (s,i) to H_0, the critical energy reproduction number R_{0E_0} is below the linearized instability reproduction number R_{0c}. By using the reduction method, the coincidence of the critical linear and nonlinear reproduction numbers has been proved and a known radius of attraction for the initial data has been given.

Here we shortly study the stability of the endemic equilibrium H_1 in the case of Dirichlet and Neumann boundary conditions. The perturbations equations to H_1 are found to be

$$s_t = \Delta s + p\Delta i - R_0\lambda s - i - R_0 si, \quad i_t = p\Delta s + \Delta i + \lambda(R_0 - 1)s + R_0 si. \tag{14}$$

It can be seen that if p is sufficiently small, $(p^2-1)\xi + p \leq 0$ or if $\lambda > (p^2-1)\xi + p$ (ξ is the *first* eigenvalue of $-\Delta$), the endemic equilibrium is linearly stable for any $R_0 > 1$.

For simplicity, we assume $p = 0$. In this case, H_1 is linearly stable for any $R_0 > 1$. By using the classical energy or the reduction method, it can be seen that, for any $R_0 > 1$, we have conditional stability with a known radius of attraction for the initial data.

In order to prove global asymptotical stability in $\mathbb{R}^2_+$ (i.e., for $s + S_e > 0$, $i + I_e > 0$), for any $R_0 > 1$, we introduce the Lyapunov function

$$E = \frac{R_0^2}{2}\|s\|^2 + R_0 \int_\Omega \left(i - I_e \ln(1 + \frac{i}{I_e}) \right) d\Omega .$$

It is easy to check that

$$\dot{E} \leq -R_0^3(s^2, \lambda + i) - R_0^2\|\nabla s\|^2 - R_0 I_e \xi \| \ln(1 + \tfrac{i}{I_e})\|^2 .$$

Since,

$$i + \lambda = i + I_e + \frac{\lambda}{R_0} > 0,$$

we have also global asymptotical stability in $\mathbb{R}^2_+$.

4 Conclusion

We have shown necessary and sufficient stability conditions for constant solutions of three systems with the reduction method: two competition models and an epidemic model with diffusion. Global stability results have been obtained with the choice of suitable Lyapunov functions, in particular cases (for instance for the symmetric May-Leonard system with Neumann boundary conditions) the linear and global nonlinear boundaries coincide.

Some open problems are the following: global stability of the endemic solution, for the epidemic model, in the case of cross-diffusion ($p \neq 0$) and for the disease free (at the moment the best results are given by the reduction method). Optimal global stability in the case of asymmetric May-Leonard system for any boundary condition.

References

[1] Anderson, R. M., Jackson, H. C., May, R. M. and Smith, A. M. (1981) Population dynamics of fox rabies in Europe, *Nature* **289** 765-771.

[2] Bellman, R. (1953) *Stability theory of differential equations*, McGraw-Hill, New-York.

[3] Beltrami, E. (1997) *Mathematics for Dynamic Modeling.* 2nd Ed., Academic Press (San Diego) .

[4] Cantrell, R. S., Cosner, C. (2003) *Spatial ecology via reaction-diffusion equations*, Wiley, Chichester, UK.

[5] Chia-Wei Chi, Sze-Bi Hsu, and Lih-Ing Wu (1998) On the asymmetric May-Leonard model of three competing species, *SIAM J. Appl. Math.* **58**, No. 1, 211–226.

[6] Flavin, J., Rionero, S. (1996) *Qualitative estimates for partial differential equations. An introduction.* CRC Press, Boca Raton, Florida.

[7] Fitzgibbon, B., Hollis, S. L. and Morgan J. J. (1997) Stability and Lyapunov functions for reaction-diffusion systems, *SIAM J. Math Anal.* **28** No. 3, 595-610.

[8] Galdi, G. P. and Straughan, B. (1985) Exchange of stabilities, symmetry and nonlinear stability, *Arch. Rational Mech. Anal.* **89**, 211-228.

[9] Hale, J. K. (1969) *Ordinary differential equations*, Wiley-Interscience (New-York).

[10] Hartman, P. (1982) *Ordinary differential equations.* Reprint of the second edition, Birkhäuser (Boston).

[11] Hastings, A. (1978) Global stability in Lotka-Volterra systems with diffusion, *J. Math. Biol.* **6** , no. 2, 163–168.

[12] Henry, D. (1981) *Geometric theory of semilinear parabolic equations.* Lecture Notes in Math. 840, Springer-Verlag (Berlin).

[13] Hirsch, M. W. and Smale, S. (1974) *Differential equations, dynamical systems, and linear algebra*, Academic Press, New York.

[14] Lombardo, S., Mulone, G., Trovato, M. (2008) Nonlinear stability in reaction-diffusion systems via optimal Lyapunov functions, *J. Math. Anal. Appl.* **342** n. 1, 461–476.

[15] May, R.M. and Leonard, W.J. (1975) Nonlinear aspects of competition between three species, *SIAM J. Appl. Math.* **29**, 243–253.

[16] Mulone, G. and Rionero, S. (1997) The rotating Bnard problem: new stability results for any Prandtl and Taylor numbers. *Cont. Mech. Thermodyn.* **9**, 347-363.

[17] Mulone, G. and Rionero, S. (2003) Necessary and Sufficient Conditions for Nonlinear Stability in the Magnetic Bènard Problem, *Arch. Rational Mech. Anal.* **166**, n. 3 197–218.

[18] Mulone G., Straughan, B. (2006) An operative method to obtain necessary and sufficient stability conditions for double diffusive convection in porous media. *ZAMM* **86**, n. 7, 507-520.

[19] Mulone, G. and Straughan, B. (2008) Nonlinear stability of multi-species ecological problems (submitted).

[20] Mulone, G., Straughan, B. and Wang W. (2007) Stability of Epidemic Models with Evolution. *Stud. Appl. Math.* **118**, 117–132.

[21] Murray, J.D. (2002) *Mathematical biology. I. An introduction.* Third edition. Interdisciplinary Applied Mathematics, 17. Springer-Verlag, New York.

[22] Murray, J.D. (2003) *Mathematical biology. II. Spatial models and biomedical applications.* Third edition. Interdisciplinary Applied Mathematics, 18. Springer-Verlag, New York.

[23] Pao, C. V. (2004) Global asymptotic stability of Lotka-Volterra competition systems with diffusion and time delays. *Nonlinear Anal. Real World Appl.* **5**, no. 1, 91–104.

[24] Payne, L. E. and Straughan, B. (1999a) Convergence, continuous dependence and decay for the Brinkman-Forchheimer equations, *Stud. Appl. Math.* **102**, 419-439.

[25] Payne, L. E. and Straughan, B. (1999b) Convergence for the equations for a Maxwell fluid, *Stud. Appl. Math.* **103**, 267-278.

[26] Quinlan, R. A. and Straughan, B. (2005) Decay bounds in a model for aggregation of microglia: application to Alzheimer's disease senile plaques. *Proc. R. Soc. Lond. Ser. A* **461**, no. 2061, 2887–2897.

[27] Rionero, S. (2006) A rigorous reduction of the L^2-stability of the solutions to a nonlinear binary reaction - diffusion system of PDE's to the stability of the solutions to a linear binary system of ODE's. *J. Mat. Anal Appl.* **319**, no. 2, 377–397.

[28] Smoller, J. (1994) *Shock waves and reaction-diffusion equations*, 2nd edn., Springer (New York).

[29] Straughan, B. (2001) A sharp nonlinear stability threshold in rotating porous convection, *Proc. Roy. Soc. London A* **457**, 87-93.

[30] Straughan, B. (2002) Sharp global nonlinear stability for temperature dependent viscosity convection, *Proc. Roy. Soc. London A* **458**, 1773-1782.

[31] Straughan, B. (2004) *The Energy Method, Stability, and Nonlinear Convection*, 2nd edn., Ser. in Appl. Math. Sci., **91**, Springer (New-York).

[32] Webber, M. and Straughan, B. (2006) Decay estimates in chemotaxis: aggregation of glia and a possible application to Alzheimer's disease senile plaques. in *"WASCOM 2005"—13th Conference on Waves and Stability in Continuous Media,* 548–552, World Sci. Publ., Hackensack, NJ.

On the Spatial Behaviour for Transversely Isotropic Plates

F. Passarella[*], *V. Zampoli*[†]

Dipartimento di Ingegneria dell'Informazione e Matematica Applicata
Università degli Studi di Salerno

Abstract

Taking into account the Mindlin model about the theory concerning a state of bending for a linear transversely isotropic plate, some results about the spatial behaviour of transient solutions are established through appropriate families of line-integral measures. It has to be underlined that results are obtained under relaxed hypotheses on the positive definiteness of elasticity tensor.

1 Introduction

Many practical applications, from building materials to electronic production, show connections with the theory about elastic plates. Several authors, such as Naghdi [1], Green and Naghdi [2], Lagnese and Lions [3], dedicate several articles to the study of properties of such particular mechanical structure.

It has to be remarked that the classical linear theory concerning thin plates (based on Kirchhoff's elastic strain-displacement relations) completely neglects the effects of transverse shear forces, cf. Lagnese and Lions [3] and Nowinski [4]. On the contrary, more refined models take into account not only the deflection of plate's middle section, but also transverse shear deformations. In particular, in the Mindlin thin elastic plate theory (cf. Mindlin [5] and Reissner [6]), the assumptions about the displacement field allow to consider separately the extensional motion and the state of bending. In this context, it is possible to remark that the theory based on the Mindlin-model is developed by Constanda [7] for the elastostatic bending of a thin plate, including the effects of transverse shear deformation; furthermore, a dynamic model for small deformations of a thin thermoelastic plate is described by Schiavone and Tait [8].

Transversely isotropic materials are characterized by 5 elastic constants and have interesting applications in civil, mechanical and aerospace engineering; their main feature is that it is possible to find an axis that allows a rotation of the material about it without changing its properties; the plane perpendicular to this axis of rotational symmetry is called plane of isotropy (consider, for example, laminates made of randomly oriented chopped fibers placed in a certain plane). Each plane containing the axis of

[*]Email: passarella@diima.unisa.it
[†]Email: vzampoli@unisa.it

rotation is a plane of symmetry and therefore transversely isotropic materials admit an infinite number of elastic symmetries.

Such materials are widely investigated by Gurtin [9], Padovani [10], Merodio and Ogden [11], Chiri]tă [12], Chiri]tă, Danescu and Ciarletta [13] and Zhang, Ding and Chen [14]. Moreover, Simmonds [15] obtains an exact 3-D Lévy-type solution for bending of an elastic slab taking into account transversal isotropy. In particular, Merodio and Ogden [11], Chiri]tă [12] and Chiri]tă, Danescu and Ciarletta [13] establish the necessary and sufficient conditions for strong ellipticity characterizing a transversely isotropic elastic solid. Passarella and Zampoli [16] reach a result similar to Brun's theorem, a uniqueness result and a Galerkin representation of solution of the field equations for a transversely isotropic plate described through the Mindlin-model and without positive definiteness assumptions on the elasticity tensor.

In the present work, the strong ellipticity condition on the elasticity tensor is taken into account, first showing the linear theory characterizing a state of bending for (bounded and unbounded) plates of Mindlin-type and then, through the time-weighted surface power function method (cf. Chiri]tă and Ciarletta [17, 18] and Ciarletta, Chiri]tă and Passarella [19]), deriving the first-order differential inequalities satisfied by a family of appropriate integral measures depending on a specific parameter. A spatial decay estimate of Saint-Venant's type with time-independent decay rate is obtained and the existence of a domain of influence is proved.

2 Basic model

The behaviour of a homogeneous transversely isotropic elastic solid that occupies at time $t = 0$ the right cylinder $\overline{B}$ of length $2h$ with the (bounded and unbounded) cross-section $\overline{\Sigma}$ and the smooth lateral boundary Π is investigated in the present paper. B and Σ denote the interiors of $\overline{B}$ and $\overline{\Sigma}$ respectively and the rectangular Cartesian coordinate frame is chosen in such a way that Ox_1x_2 is the middle plane of the plate and thus its faces are situated at $x_3 = \pm h$. Moreover, Σ is supposed to be a simply connected region, L is the boundary of Σ and $h \ll \text{diam}\ \Sigma$.

As usual, Latin subscripts (unless otherwise specified) are understood to range over the integers $\{1, 2, 3\}$, whereas Greek subscripts are confined to the range $\{1, 2\}$; summation over repeated subscripts is implied. Superposed dots or subscripts preceded by a comma mean partial derivative with respect to the time or the corresponding Cartesian variables.

Moreover, regularity questions are simply disregarded, understanding a degree of smoothness sufficient to ensure analysis to be valid. According to the linear theory of elastodynamics, the field equations are

$$\begin{aligned}
&\text{equations of motion} && t_{ji,j} + f_i = \varrho \ddot{u}_i, \qquad \text{on } B \times (0, \infty) \\
&\text{stress–strain relation} && t_{ij} = C_{ijkl} e_{kl} \\
&\text{strain–displacement relation} && e_{ij} = \frac{1}{2}(u_{i,j} + u_{j,i})
\end{aligned} \tag{1}$$

where u_i are the components of the displacement vector, t_{ij} are the components of the stress tensor, ρ is the reference mass density and C_{ijkl} are the components of

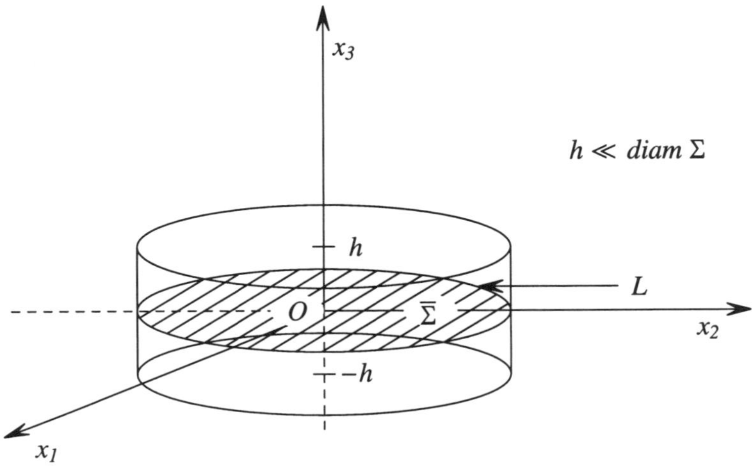

the symmetric elasticity tensor. According to Gurtin [9], the direction of transversal isotropy is assumed to be the one indicated by the x_3 coordinate axis and

$$c_{ij} = C_{iijj} = c_{ji}, \qquad i,j \text{ (not summed)} \in \{1,2,3\}, \qquad c_{11} = c_{22}, \qquad c_{13} = c_{23},$$

$$c_{44} = c_{55} = C_{2323} = C_{1313}, \qquad c_{66} = C_{1212} = \frac{c_{11} - c_{12}}{2}. \tag{2}$$

These are the only non-zero components of C_{ijkl} apart from the terms obtained using symmetries (2).

The class of transversely isotropic materials having a strongly elliptic elasticity tensor is characterized by

$$C_{ijkl} m_i m_k n_j n_l > 0 \qquad \text{for every non-zero vector } m_i, n_i;$$

the strong ellipticity condition becomes

$$c_{66}\left(n_1 m_2 - n_2 m_1\right)^2 + c_{11}\left(n_1 m_1 + n_2 m_2\right)^2 + 2\left(c_{13} + c_{44}\right)\left(n_1 m_1 + n_2 m_2\right) n_3 m_3 +$$

$$+ c_{33} n_3^2 m_3^2 + c_{44}\left(n_3^2 m_1^2 + n_1^2 m_3^2 + n_3^2 m_2^2 + n_2^2 m_3^2\right) > 0$$

or equivalently (see e.g. Merodio and Ogden [11] and Chiri]tă [12])

$$c_{11} > 0, \qquad c_{11} > c_{12}, \qquad c_{33} > 0, \qquad c_{44} > 0, \qquad |c_{13} + c_{44}| < c_{44} + \sqrt{c_{11} c_{33}}. \tag{3}$$

Gurtin [9] proves that $\mathbf{C}$ is positive definite if

$$C_{ijkl} \phi_{ij} \phi_{kl} > 0 \qquad \text{for every non-zero symmetric tensor } \phi_{ij}$$

and then it is strongly elliptic too.

A state of bending for an elastic plate is characterized by the following relations

$$u_\alpha(x_1,x_2,x_3,t) = -u_\alpha(x_1,x_2,-x_3,t), \qquad u_3(x_1,x_2,x_3,t) = u_3(x_1,x_2,-x_3,t)$$

for every $(x_1,x_2,x_3) \in B$, $t \in [0,\infty)$; moreover, we assume that the body loads obey to the relations

$$f_\alpha(x_1,x_2,x_3,t) = -f_\alpha(x_1,x_2,-x_3,t), \quad f_3(x_1,x_2,x_3,t) = f_3(x_1,x_2,-x_3,t).$$

In the present paper, the attention is restricted to a state of bending characterized by

$$u_\alpha(x_1,x_2,x_3,t) = x_3 v_\alpha(x_1,x_2,t), \qquad u_3(x_1,x_2,x_3,t) = w(x_1,x_2,t). \tag{4}$$

Then, using the following notations

$$\begin{aligned}
M_{ij}(x_1,x_2,t) &= \frac{3}{2h^3}\int_{-h}^{h} x_3 t_{ij}(x_1,x_2,x_3,t)dx_3,\\
\tau_{ij}(x_1,x_2,t) &= \frac{1}{2h}\int_{-h}^{h} t_{ij}\,(x_1,x_2,x_3,t)dx_3,\\
F(x_1,x_2,t) &= \frac{1}{2h}\int_{-h}^{h} f_3(x_1,x_2,x_3,t)\,dx_3 + \frac{1}{h}t_{33}(x_1,x_2,h,t),\\
H_\alpha(x_1,x_2,t) &= \frac{1}{2h}\int_{-h}^{h} x_3 f_\alpha(x_1,x_2,x_3,t)\,dx_3 + t_{3\alpha}(x_1,x_2,h,t),
\end{aligned}$$

it is easy to obtain $M_{\alpha 3} = 0$, $\tau_{\alpha\beta} = 0$, $\tau_{33} = 0$ since $t_{\alpha\beta}$, t_{33} are odd functions and $t_{\alpha 3}$ is even with respect to x_3. Thus, the basic equations consist in the equations of motion

$$\frac{h^2}{3}M_{\beta\alpha,\beta} - \tau_{\alpha 3} + H_\alpha = \rho\frac{h^2}{3}\ddot{v}_\alpha, \qquad \tau_{\beta 3,\beta} + F = \rho\ddot{w} \qquad \text{on } \overline{\Sigma}\times(0,\infty), \tag{5}$$

the constitutive equations

$$\begin{aligned}
M_{11} &= c_{11}\epsilon_{11} + c_{12}\epsilon_{22}, & M_{12} &= M_{21} = (c_{11}-c_{12})\epsilon_{12},\\
M_{22} &= c_{12}\epsilon_{11} + c_{11}\epsilon_{22}, & \tau_{\alpha 3} &= \tau_{3\alpha} = 2c_{44}\xi_\alpha
\end{aligned} \tag{6}$$

and the geometrical equations

$$\epsilon_{\alpha\beta} = \frac{1}{2}(v_{\alpha,\beta} + v_{\beta,\alpha}), \qquad \xi_\alpha = \frac{1}{2}\left(w_{,\alpha} + v_\alpha\right). \tag{7}$$

In what follows, we assume the hypotheses

$$H_\alpha(x_1,x_2,t) = 0, \qquad F(x_1,x_2,t) = 0,$$

and we rewrite the system (5) - (7) in terms of $\{v_\alpha, w\}$ as

$$\begin{aligned}
&\frac{h^2}{3}\frac{c_{11}-c_{12}}{2}v_{\alpha,\beta\beta} + \frac{h^2}{3}\frac{c_{11}+c_{12}}{2}v_{\beta,\beta\alpha} - c_{44}(w_{,\alpha}+v_\alpha) = \rho\frac{h^2}{3}\ddot{v}_\alpha,\\
&c_{44}w_{,\beta\beta} + c_{44}v_{\beta,\beta} = \rho\ddot{w}.
\end{aligned} \tag{8}$$

It is now necessary to consider the problem $\mathcal{P}$ defined by (8), the homogeneous initial conditions

$$v_\alpha(x_1,x_2,0)=0, \quad w(x_1,x_2,0)=0, \quad \dot{v}_\alpha(x_1,x_2,0)=0, \quad \dot{w}(x_1,x_2,0)=0, \quad \text{on } \overline{\Sigma} \quad (9)$$

and the boundary conditions (Dirichlet problem)

$$v_\alpha=\breve{v}_\alpha, \qquad w=\breve{w}, \qquad \text{on } L\times[0,+\infty), \quad (10)$$

where $\breve{v}_\alpha$ and $\breve{w}$ are assigned fields representing external data $\mathcal{D}$ of the problem in concern.

It has now to be remarked that, cf. Passarella and Zampoli [16], if and only if $c_{44}>0$, $c_{11}>0$, $|c_{12}|<c_{11}$, the internal energy density of a transversely isotropic elastic plate (considering a state of bending) is a positive definite quadratic form. It is interesting to notice that, from strong ellipticity conditions (3) it is $c_{12}<c_{11}$ and thus a relaxation of previous restrictions on constitutive coefficients exists.

3 Spatial behaviour of transient solutions

In this section, results describing the spatial behaviour of solutions of given data on the interval $[0,T]$ are given under the strong ellipticity condition on the elasticity tensor.

The system (8) can be rewritten as follows

$$\frac{h^2}{3}\widehat{M}_{\beta\alpha,\beta}-\tau_{\alpha 3}=\rho\frac{h^2}{3}\ddot{v}_\alpha, \qquad \tau_{\beta 3,\beta}=\rho\ddot{w} \qquad \text{on } \overline{\Sigma}\times(0,\infty), \quad (11)$$

where $\widehat{M}_{\beta\alpha}$ are families of tensors depending on a parameter κ

$$\begin{aligned}
\widehat{M}_{11}&=c_{11}v_{1,1}+\frac{c_{11}-\kappa}{2}v_{2,2}, & \widehat{M}_{12}&=\frac{c_{12}+\kappa}{2}v_{1,2}+\frac{c_{11}-c_{12}}{2}v_{2,1},\\
\widehat{M}_{21}&=\frac{c_{11}-c_{12}}{2}v_{1,2}+\frac{c_{12}+\kappa}{2}v_{2,1}, & \widehat{M}_{22}&=\frac{c_{11}-\kappa}{2}v_{1,1}+c_{11}v_{2,2}.
\end{aligned} \quad (12)$$

To give an example, if $\kappa=-c_{12}$ then $\widehat{M}_{\beta\alpha}=\dfrac{c_{11}-c_{12}}{2}v_{\alpha,\beta}+\dfrac{c_{11}+c_{12}}{2}\delta_{\alpha\beta}v_{\tau,\tau}$, and if $\kappa=c_{11}$ then $\widehat{M}_{\beta\alpha}=\dfrac{c_{11}-c_{12}}{2}v_{\alpha,\beta}+\dfrac{c_{11}+c_{12}}{2}v_{\beta,\alpha}$.

It is necessary to define now the support D_T^* of external data $\mathcal{D}$ on the time interval $[0,T]$ and, for convenience, D_T^* is assumed to be a bounded regular region of $\overline{\Sigma}$. The following sets have also to be considered

$$D_r=\left\{\mathbf{x}\in\overline{\Sigma}:\ D_T^*\cap\overline{S(\mathbf{x},r)}\neq\emptyset\right\}, \qquad \Sigma_r=\overline{\Sigma}\backslash D_r, \qquad \Sigma(r_1,r_2)=\Sigma_{r_2}\backslash\Sigma_{r_1},$$

where $r\geq 0, r_2\leq r_1$ and $\overline{S(\mathbf{x},r)}$ is the closed ball with radius r and center at $\mathbf{x}$. Then, L_r is the subcurve of $\partial\Sigma_r$ contained inside $\overline{\Sigma}$ and whose outward unit normal vector $\mathbf{n}$ is forwarded to the exterior of D_r. The following function has to be defined for any positive parameter λ

$$\mathcal{I}_\kappa(r,t)=-\int_0^t\int_{L_r}e^{-\lambda s}\left[\tau_{\beta 3}(s)\dot{w}(s)+\frac{h^2}{3}\widehat{M}_{\beta\alpha}(s)\dot{v}_\alpha(s)\right]n_\beta\,dl\,ds, \quad r\geq 0, \quad t\in[0,T] \quad (13)$$

and through Eqs. (11), (12) it is easy to show that

$$\left[\tau_{\beta 3}\dot{w} + \frac{h^2}{3}\widehat{M}_{\beta\alpha}\dot{v}_\alpha\right]_{,\beta} = \frac{\partial}{\partial s}\left[\mathcal{T} + \mathcal{W}\right], \tag{14}$$

where

$$\begin{aligned} &\mathcal{T} = \frac{\rho}{2}\left(\dot{w}^2 + \frac{h^2}{3}\dot{v}_\alpha\dot{v}_\alpha\right), \qquad \mathcal{W} = \frac{1}{2}\left(\mathcal{W}_0 + \frac{h^2}{3}\mathcal{W}_1 + \frac{h^2}{3}\mathcal{W}_2\right), \\ &\mathcal{W}_0 = 4c_{44}(\xi_1^2 + \xi_2^2), \qquad \mathcal{W}_1 = c_{11}\left(v_{1,1}^2 + v_{2,2}^2\right) + (c_{11} - \kappa)\, v_{1,1}v_{2,2}, \\ &\mathcal{W}_2 = \frac{c_{11} - c_{12}}{2}\left(v_{2,1}^2 + v_{1,2}^2\right) + (c_{12} + \kappa)\, v_{2,1}v_{1,2}. \end{aligned} \tag{15}$$

In order to write, for $r_2 < r_1$,

$$\begin{aligned} &\mathcal{I}_\kappa(r_1, t) - \mathcal{I}_\kappa(r_2, t) = -\int_0^t \int_{\Sigma(r_1,r_2)} e^{-\lambda s}\frac{\partial}{\partial s}\left[\mathcal{T}(s) + \mathcal{W}(s)\right] d\sigma\, ds = \\ &= -\int_{\Sigma(r_1,r_2)} e^{-\lambda t}\left[\mathcal{T}(t) + \mathcal{W}(t)\right] d\sigma - \lambda\int_0^t \int_{\Sigma(r_1,r_2)} e^{-\lambda s}\left[\mathcal{T}(s) + \mathcal{W}(s)\right] d\sigma\, ds, \end{aligned} \tag{16}$$

the definitions of Σ_r and $\mathcal{I}_\kappa(r,t)$, the divergence theorem and Eq. (14) have to be used. It can then be proved that function $\mathcal{I}_\kappa(r,t)$ satisfies some properties.

Theorem 1. Let ρ be greater than zero, the strong ellipticity condition (3) be valid and κ satisfy the relation

$$-c_{11} < \kappa < \min\left\{3c_{11}, c_{11} - 2c_{12}\right\}. \tag{17}$$

Corresponding to a solution $\{v_\alpha, w\}$ of initial-boundary value problem $\mathcal{P}$ and called $\widehat{D}_T$ the bounded support of external data $\mathcal{D}$ on the time interval $[0, T]$, it is possible to prove that

1) $\mathcal{I}_\kappa(r,t)$ is a continuous differentiable function on $r \geq 0$ and

$$\frac{\partial \mathcal{I}_\kappa}{\partial t}(r,t) = -\int_{L_r} e^{-\lambda t}\left[\tau_{\beta 3}(t)\dot{w}(t) + \frac{h^2}{3}\widehat{M}_{\beta\alpha}(t)\dot{v}_\alpha(t)\right] n_\beta\, dl, \tag{18}$$

$$\frac{\partial \mathcal{I}_\kappa}{\partial r}(r,t) = -\int_{L_r} e^{-\lambda t}\left[\mathcal{T}(t) + \mathcal{W}(t)\right] dl - \lambda\int_0^t \int_{L_r} e^{-\lambda s}\left[\mathcal{T}(s) + \mathcal{W}(s)\right] dl\, ds. \tag{19}$$

2) $\mathcal{I}_\kappa(r,t)$ is a non-increasing function with respect to r, i.e.

$$\mathcal{I}_\kappa(r_1, t) \leq \mathcal{I}_\kappa(r_2, t) \qquad r_2 < r_1. \tag{20}$$

3) $\mathcal{I}_\kappa(r,t)$ satisfies the following first-order differential inequalities

$$|\mathcal{I}_\kappa(r,t)| + \frac{c_\kappa}{\lambda}\frac{\partial \mathcal{I}_\kappa}{\partial r}(r,t) \leq 0 \quad \text{and} \quad \left|\frac{\partial \mathcal{I}_\kappa}{\partial t}(r,t)\right| + c_\kappa\frac{\partial \mathcal{I}_\kappa}{\partial r}(r,t) \leq 0, \tag{21}$$

where

$$c_\kappa = \sqrt{\frac{\eta_\kappa}{\rho}} \qquad \text{with} \quad \eta_\kappa = \max\left\{c_{44}, \frac{c_{11} + \kappa}{2}, \frac{3c_{11} - \kappa}{2}, \frac{c_{11} - 2c_{12} - \kappa}{2}\right\}. \tag{22}$$

4) $\mathcal{I}_\kappa(r,t)$ is a non-negative function and

$$0 \le \mathcal{I}_\kappa(r,t) = \int_{\Sigma_r} e^{-\lambda t}\left[\mathcal{T}(t)+\mathcal{W}(t)\right] d\sigma + \lambda \int_0^t \int_{\Sigma_r} e^{-\lambda s}\left[\mathcal{T}(s)+\mathcal{W}(s)\right] d\sigma\, ds. \quad (23)$$

Proof. From a direct differentiation of Eq. (13) with respect to the variable t, Eq. (18) is obtained, while from Eq. (16) and through a differentiation with respect to r, it is possible to write Eq. (19). The kinetic energy $\mathcal{T}$ and the quadratic form $\mathcal{W}$ are positive definite if $\rho > 0$ and hypotheses (3), (17) hold, thus Eq. (16) implies that $\mathcal{I}_\kappa(r,t)$ is a non-increasing function with respect to r.

$\mathcal{I}_\kappa(r,t)$ and $\dfrac{\partial \mathcal{I}_\kappa}{\partial t}(r,t)$ can be estimated in terms of $\dfrac{\partial \mathcal{I}_\kappa}{\partial r}(r,t)$. Denoted by $\mathcal{A}_\alpha$ the matrix associated to the quadratic form $\mathcal{W}_\alpha$ and considered the functional $\mathcal{F}[\mathcal{A}_\alpha;\varphi,\gamma] \equiv \varphi\cdot\mathcal{A}_\alpha\gamma$ for the variables $\varphi = \{\varphi_1,\varphi_2\}$ and $\gamma = \{\gamma_1,\gamma_2\}$, hypotheses (3), (17) imply

$$k_m^{(\alpha)}(\gamma_1^2+\gamma_2^2) \le \mathcal{F}[\mathcal{A}_\alpha;\gamma,\gamma] \le k_M^{(\alpha)}(\gamma_1^2+\gamma_2^2), \quad (24)$$

where

$$\begin{aligned}
k_m^{(1)} &= \min\left\{\frac{c_{11}+\kappa}{2}, \frac{3c_{11}-\kappa}{2}\right\}, & k_m^{(2)} &= \min\left\{\frac{c_{11}+\kappa}{2}, \frac{c_{11}-2c_{12}-\kappa}{2}\right\}, \\
k_M^{(1)} &= \max\left\{\frac{c_{11}+\kappa}{2}, \frac{3c_{11}-\kappa}{2}\right\}, & k_M^{(2)} &= \max\left\{\frac{c_{11}+\kappa}{2}, \frac{c_{11}-2c_{12}-\kappa}{2}\right\}.
\end{aligned} \quad (25)$$

Moreover, through Schwarz inequality and Eqs. (12), (15),

$$\mathcal{F}[\mathcal{A}_\alpha;\varphi,\gamma] \le [\mathcal{F}[\mathcal{A}_\alpha;\varphi,\varphi]]^{1/2}[\mathcal{F}[\mathcal{A}_\alpha;\gamma,\gamma]]^{1/2} \quad (26)$$

and

$$\begin{aligned}
\mathcal{F}\left[\mathcal{A}_1;\widehat{\mathbf{M}}^{(1)},\widehat{\gamma}^{(1)}\right] &= \widehat{M}_{11}^2+\widehat{M}_{22}^2, & \mathcal{F}\left[\mathcal{A}_1;\widehat{\gamma}^{(1)},\widehat{\gamma}^{(1)}\right] &= \mathcal{W}_1, \\
\mathcal{F}\left[\mathcal{A}_2;\widehat{\mathbf{M}}^{(2)},\widehat{\gamma}^{(2)}\right] &= \widehat{M}_{21}^2+\widehat{M}_{12}^2, & \mathcal{F}\left[\mathcal{A}_2;\widehat{\gamma}^{(2)},\widehat{\gamma}^{(2)}\right] &= \mathcal{W}_2,
\end{aligned} \quad (27)$$

where

$$\widehat{\mathbf{M}}^{(1)} = \left\{\widehat{M}_{11},\widehat{M}_{22}\right\}, \qquad \widehat{\gamma}^{(1)} = \{v_{1,1},v_{2,2}\}, \qquad \widehat{\mathbf{M}}^{(2)} = \left\{\widehat{M}_{21},\widehat{M}_{12}\right\}, \qquad \widehat{\gamma}^{(2)} = \{v_{1,2},v_{2,1}$$

Collecting Eqs. (15), (24) - (27), it is easy to deduce that

$$\begin{aligned}
&\tau_{\beta 3}\tau_{\beta 3} = c_{44}\mathcal{W}_0, \\
&\widehat{M}_{11}^2+\widehat{M}_{22}^2 \le \mathcal{F}^{1/2}\left[\mathcal{A}_1;\widehat{\mathbf{M}}^{(1)},\widehat{\mathbf{M}}^{(1)}\right]\mathcal{W}_1^{1/2} \le \left[k_M^{(1)}\left(\widehat{M}_{11}^2+\widehat{M}_{22}^2\right)\right]^{1/2}\mathcal{W}_1^{1/2}, \\
&\widehat{M}_{21}^2+\widehat{M}_{12}^2 \le \mathcal{F}^{1/2}\left[\mathcal{A}_2;\widehat{\mathbf{M}}^{(2)},\widehat{\mathbf{M}}^{(2)}\right]\mathcal{W}_2^{1/2} \le \left[k_M^{(2)}\left(\widehat{M}_{21}^2+\widehat{M}_{12}^2\right)\right]^{1/2}\mathcal{W}_2^{1/2}
\end{aligned} \quad (28)$$

and, consequently, to conclude that

$$\tau_{\beta 3}\tau_{\beta 3} + \frac{h^2}{3}\widehat{M}_{\beta\alpha}\widehat{M}_{\beta\alpha} \le c_{44}\mathcal{W}_0 + \frac{h^2 k_M^{(1)}}{3}\mathcal{W}_1 + \frac{h^2 k_M^{(2)}}{3}\mathcal{W}_2 \le 2\eta_\kappa\mathcal{W}, \quad (29)$$

where $\eta_\kappa = \max\left\{c_{44}, k_M^{(1)}, k_M^{(2)}\right\}$ is given by Eq. (22). Using Cauchy-Schwarz and arithmetic-geometric mean inequalities, it is

$$\left|\left[\tau_{\beta 3}\dot{w} + \frac{h^2}{3}\widehat{M}_{\beta\alpha}\dot{v}_\alpha\right] n_\beta\right| \leq \frac{\epsilon\rho}{2}\left[\dot{w}^2 + \frac{h^2}{3}\dot{v}_\alpha\dot{v}_\alpha\right] + \frac{1}{2\epsilon\rho}\left[\tau_{\beta 3}\tau_{\beta 3} + \frac{h^2}{3}\widehat{M}_{\beta\alpha}\widehat{M}_{\beta\alpha}\right]. \quad (30)$$

Moreover, estimates (29), (30) should be used into Eqs. (13) and (18) in order to obtain

$$|\mathcal{I}_\kappa(r,t)| \leq \epsilon \int_0^t \int_{L_r} e^{-\lambda s}\left[\mathcal{T}(s) + \frac{\eta_\kappa}{\epsilon^2\rho}\mathcal{W}(s)\right] dl\, ds, \quad (31)$$

$$\left|\frac{\partial \mathcal{I}_\kappa}{\partial t}(r,t)\right| \leq \epsilon \int_{L_r} e^{-\lambda t}\left[\mathcal{T}(t) + \frac{\eta_\kappa}{\epsilon^2\rho}\mathcal{W}(t)\right] dl\,. \quad (32)$$

Remembering Eq. (19) and setting $\epsilon = c_\kappa = \sqrt{\dfrac{\eta_\kappa}{\rho}}$ into (31), (32), Eqs. (21) can be verified.

If the plate is bounded, then r ranges in $\left[0, \ell = \max_{\mathbf{x}\in\overline{\Sigma}}\left\{\min_{\mathbf{y}\in D_T^*}\sqrt{(x_i - y_i)(x_i - y_i)}\right\}\right]$. Considering that

$$\mathcal{I}_\kappa(r,0) = 0, \qquad \mathcal{I}_\kappa(\ell,t) = 0, \qquad r \in [0,\ell], \qquad t \in [0,T]$$

and setting $r_1 = \ell$, $r_2 = r \in [0,\ell]$ into relations (16), (20), one can arrive to Eq. (23). On the other hand, if the plate is unbounded, then the variable r ranges in $[0,\infty)$. Eq. $(21)_2$ can be rewritten as follows

$$\frac{\partial \mathcal{I}_\kappa}{\partial t}(r,t) + c_\kappa\,\frac{\partial \mathcal{I}_\kappa}{\partial r}(r,t) \leq 0, \qquad -\frac{\partial \mathcal{I}_\kappa}{\partial t}(r,t) + c_\kappa\,\frac{\partial \mathcal{I}_\kappa}{\partial r}(r,t) \leq 0. \quad (33)$$

Choosing a pair (r_0, t_0) such that $t_0 \in [0,T]$ and $r_0 \geq c_\kappa t_0$ and putting $t = t_0 + \dfrac{r - r_0}{c_\kappa}$ into inequality $(33)_1$, it is possible to obtain

$$\frac{d}{dr}\left[\mathcal{I}_\kappa\left(r, t_0 + \frac{r - r_0}{c_\kappa}\right)\right] \leq 0 \quad \Rightarrow \quad \mathcal{I}_\kappa\left(r_1, t_0 + \frac{r_1 - r_0}{c_\kappa}\right) \leq \mathcal{I}_\kappa\left(r_2, t_0 + \frac{r_2 - r_0}{c_\kappa}\right), \quad r_1 \geq r_2,$$

so that, considering $r_1 = r_0$ and $r_2 = r_0 - c_\kappa t_0$, one can deduce

$$\mathcal{I}_\kappa(r_0, t_0) \leq \mathcal{I}_\kappa(r_0 - c_\kappa t_0, 0) = 0. \quad (34)$$

Similarly, setting $t = t_0 - \dfrac{r - r_0}{c_\kappa}$ in inequality $(33)_2$, then

$$\frac{d}{dr}\left[\mathcal{I}_\kappa\left(r, t_0 - \frac{r - r_0}{c_\kappa}\right)\right] \leq 0 \quad \Rightarrow \quad \mathcal{I}_\kappa\left(r_1, t_0 - \frac{r_1 - r_0}{c_\kappa}\right) \leq \mathcal{I}_\kappa\left(r_2, t_0 - \frac{r_2 - r_0}{c_\kappa}\right), \quad r_1 \geq r_2$$

and thus, for $r_1 = r_0 + c_\kappa t_0$ and $r_2 = r_0$,

$$\mathcal{I}_\kappa(r_0, t_0) \geq \mathcal{I}_\kappa(r_0 + c_\kappa t_0, 0) = 0. \quad (35)$$

The following relation can be deduced from (34) and (35)

$$\mathcal{I}_\kappa(\infty, t) \equiv \lim_{r_0 \to \infty} \mathcal{I}_\kappa(r_0, t_0) = 0. \quad (36)$$

Through Eqs. (16), (20), (36), it is possible to arrive to inequality (23) and the proof is complete. ∎

The results obtained up to now allow to formulate the following theorem.

Theorem 2. Let hypotheses of previous theorem be still valid. Then, for each fixed $t \in [0, T]$, the following results are valid.

1. *Spatial decay*:

$$\mathcal{I}_\kappa(r,t) \leq \exp\left(-\frac{\lambda}{c_\kappa} r\right) \mathcal{I}_\kappa(0,t), \qquad 0 \leq r < c_\kappa t. \tag{37}$$

2. *Domain of influence*:

$$v_\alpha = w = 0, \qquad \text{on } \Sigma_r \times [0,T], \quad c_\kappa t \leq r. \tag{38}$$

Proof. Since $\mathcal{I}_\kappa(r,t)$ is a non-negative function, then from Eq. $(21)_1$

$$\frac{\partial}{\partial r}\left[\exp\left(\frac{\lambda}{c_\kappa} r\right) \mathcal{I}_\kappa(r,t)\right] \leq 0 \quad \Longrightarrow \quad \mathcal{I}_\kappa(r,t) \leq \exp\left(-\frac{\lambda}{c_\kappa} r\right) \mathcal{I}_\kappa(0,t).$$

Now, putting $r = c_\kappa t$ into $(33)_1$, one can write

$$\frac{d}{dt}[\mathcal{I}_\kappa(c_\kappa t, t)] \leq 0 \quad \Rightarrow \quad \mathcal{I}_\kappa(c_\kappa t, t) \leq \mathcal{I}_\kappa(0,0) = 0, \quad 0 \leq t. \tag{39}$$

On the other hand, from Eqs. (20), (23), it is true that

$$0 \leq \mathcal{I}_\kappa(r,t) \leq \mathcal{I}_\kappa(c_\kappa t, t), \qquad r \geq c_\kappa t. \tag{40}$$

Thus, Eqs. (39), (40) imply

$$\mathcal{I}_\kappa(r,t) = 0, \qquad \text{for } r \geq c_\kappa t. \tag{41}$$

Obviously, Eq. (38) follows from Eqs. (23), (41) and from the positive definiteness of $\mathcal{T}$ and $\mathcal{W}$. ∎

According to Gurtin [9], the domain of influence of external data at time T can be depicted as the set of points of $\overline{\Sigma}$ that can be reached by signals propagating from the support $\hat{D}_T$ on the time interval $[0, T]$, with speeds equal to or less than the maximum speed of propagation

$$c = \sqrt{\frac{\eta}{\rho}} \qquad \text{with } \eta = \inf_\kappa \eta_\kappa$$

where $\kappa \in \,]-c_{11}, \min\{3c_{11}, c_{11} - 2c_{12}\}[$; in particular, $|c_{12}| < c_{11} \Rightarrow \kappa \in \,]-c_{11}, c_{11} - 2c_{12}[$, while $c_{12} \leq -c_{11} \;\Rightarrow \kappa \in \,]-c_{11}, 3c_{11}[$.

References

[1] P. M. Naghdi, *The Theory of Shells and Plates*, vol. VI a/2 of "Handbuch der Physik" (Edited by C. Truesdell), Springer-Verlag, Berlin, Heidelberg, New York, 1972.

[2] A. E. Green, P. M. Naghdi, *The linear theory of an elastic Cosserat plate*, Proc. Cambridge Phil. Soc., **63**, 537-550, 1967.

[3] J. E. Lagnese, J. L. Lions, *Modelling, analysis and control of thin plates*, Collection RMA, vol. 6, Masson, Paris, 1988.

[4] J. L. Nowinski, *Theory of thermoelasticity with applications*, Sijthoff & Noordoff, The Netherlands, 1978.

[5] R. D. Mindlin, *Influence of rotatory inertia and shear on flexural motions of isotropic elastic plates*, J. Appl. Mech., **18**, 31-38, 1951.

[6] E. Reissner, *On the theory of bending of elastic plates*, J. Math. Phys., **23,** 184-191, 1944.

[7] C. Constanda, *A mathematical analysis of bending of thin plates with transverse shear deformation*, Longman Scientific and Technical, Essex, 1990.

[8] P. Schiavone, R. J. Tait, *Thermal effects in Mindlin-type plates*, Quart. J. Mech. Appl. Math., **46**, 27-39, 1993.

[9] M. E. Gurtin, *Linear theory of elasticity*, in Handbuch der Physik (Flügge S. ed.), Springer-Verlag, Berlin, vol. VIa/2, 1-295, 1972.

[10] C. Padovani, *Strong ellipticity of transversely isotropic elasticity tensors*, Meccanica, **37**, 515-525, 2002.

[11] J. Merodio, R. W. Ogden, *A note on strong ellipticity for transversely isotropic linearly elastic solids,* Quart. J. Mec. Appl. Math., **56**, 589-591, 2003.

[12] S. Chiri]tă, *On the strong ellipticity condition for transversely isotropic linearly elastic solids,* An. St. Univ. "Al I. Cuza" Iasi, Matematica, **52**, 245-250, 2006.

[13] S. Chiri]tă, A. Danescu, M. Ciarletta, *On the strong ellipticity of the anisotropic linearly elastic materials*, Journal of Elasticity, **87**, n. 1, 1-27, 2007.

[14] L. Zhang, H. Ding, H. Chen, *Elasticity of transversely isotropic materials*, Solid mechanics and its applications, Vol. 126, Springer, Dordrecht, 2006.

[15] J. G. Simmonds, *Exact Lévy-type solution for the plate bending exist for transversely isotropic but not for general monoclinic materials*, J. Elasticity, **75**, 49-56, 2004.

[16] F. Passarella, V. Zampoli, *Some results concerning the state of bending for transversely isotropic plates*, pre-print 2007.

[17] S. Chiri]tă, M. Ciarletta, *Time-weighted surface power function method for the study of spatial behaviour in dynamics of continua*, European Journal of Mechanics, A/Solids, **18**, 915-933, 1999.

[18] S. Chiri]tă, M. Ciarletta, *Some further growth and decay results in linear thermoelastodynamics*, J. Thermal Stresses, **26**, 889-903, 2003.

[19] M. Ciarletta, S. Chiri]tă, F. Passarella, *Some results on the spatial behaviour in linear porous elasticity*, Arch. Mech., **57**, 1, 43-65, 2005.

Liapunov Functionals for the Coincidence between the First and the Second Liapunov Stability Methods

*S. Rionero**

Department of Mathematics and Applications *R. Caccioppoli*
University of Naples *Federico II*,
via Cinzia, 80126 Naples, Italy.

Abstract

A methodology for obtaining Liapunov functionals such that – via the Liapunov Direct Method – *give the same result of the first Liapunov Approximation Method.*

1 Introduction

Let us consider the autonomous system

$$\frac{d\mathbf{x}}{dt} = A\mathbf{x} + B(\mathbf{x}) \tag{1}$$

with

$$A = \begin{pmatrix} a_{11} & \dots & a_{1n} \\ \cdots & \cdots & \cdots \\ a_{n1} & \dots & a_{nn} \end{pmatrix}, \mathbf{x} = \begin{pmatrix} x_1 \\ x_2 \\ .. \\ x_n \end{pmatrix}, B = \begin{pmatrix} B_1 \\ B_2 \\ .. \\ B_n \end{pmatrix}, \tag{2}$$

B nonlinear function such that

$$B(\mathbf{x}) = o(x_1^2 + x_2^2 + \dots + x_n^2), \tag{3}$$

and a_{ij} real constants, $(i, j = 1, 2, .., n)$.

As it is well known [1], the following theorems hold.

Theorem 1 (Liapunov first Approximation Method). *If the eigenvalues of A are real negative or have negative real part then $\mathbf{x} = 0$ is linearly asymptotically exponentially stable and is also locally nonlinearly asymptotically exponentially stable. If one eigenvalue is positive or has positive real part, then $\mathbf{x} = 0$ is unstable.*

*Email: rionero@unina.it

Theorem 2 (Liapunov Direct Method). *Let $B(0,R)$ be a ball of the phase space of radius $R>0$ centered at O and let $V(x_1,x_2,...,x_n) \in C^1(B)$ be a positive definite function in B. Then, denoting by $\dot{V} = \nabla V \cdot (A\mathbf{x} + B(\mathbf{x}))$ the temporal derivative of V along the solutions it turns out that*

i) $\dot{V} \leq 0$ *in* $B(0,R)$ *implies stability of* $\mathbf{x}=0$*;*

ii) $\dot{V} < 0$ *in* $B(0,R)$ *implies (local) asymptotic stability of* $\mathbf{x}=0$*;*

iii) $\dot{V} > 0$ *in* $B(0,R)$ *implies instability of* $\mathbf{x}=0$.

The function V is called Liapunov function. As concerns the instability condition iii), we recall that has been weakened by other authors successively. We recall here the following instability theorem [2].

Theorem 3. *Let $V \in C^1(B(O,R))$ with $V(O)=0$. If $\dot{V}$ is definite positive and V assumes positive values in any ball $B(O,r)$, $r<R$, then $\mathbf{x}=0$ is unstable.*

We remark that the power of the Direct Method is to give *without requiring the knowledge of the eigenvalues of* A, either conditions sufficient for the stability or conditions sufficient for the instability. But does not give any criterion for obtaining functions V, *linked to the invariants of* A, which (in view of theorems 2-3) give the same results of theorem 1. Our aim is to give this criterion.

In Section 2, we introduce the methodology for obtaining the aforesaid functionals V by means of 3 steps. Section 3 is devoted to the case $n=3$.

2 Methodology for obtaining Liapunov functions linked to the A invariants

Step one

Introduction of a linear transformation

$$x_i = \sum_{j=1}^{n} \alpha_{ij}\xi_j \tag{4}$$

such that (1) is reduced to

$$\begin{cases} \dfrac{d\xi_1}{dt} = b_{1n}\xi_n + B_1^*(\xi_1,\xi_2,...,\xi_n) \\ \dfrac{d\xi_2}{dt} = \xi_1 + b_{2n}\xi_n + B_2^*(\xi_1,\xi_2,...,\xi_n) \\ \dfrac{d\xi_3}{dt} = \xi_2 + b_{3n}\xi_n + B_3^*(\xi_1,\xi_2,...,\xi_n) \\ \dots\dots\dots\dots\dots\dots\dots\dots\dots\dots \\ \dfrac{d\xi_n}{dt} = \xi_{n-1} + b_{nn}\xi_n + B_n^*(\xi_1,\xi_2,...,\xi_n)\,. \end{cases} \tag{5}$$

Then it follows that

$$I_n = b_{1n},\ I_{n-1} = -b_{2n},\ I_{n-2} = b_{3n},\ ...(-1)^{n-1}b_n = I_1 \tag{6}$$

Step two

Since our aim is to look for obtaining – via Direct Method – the stability conditions of the first approximation method, one can consider (under the assumption (3)) only the linear stability.

Linearizing (5), one obtains

$$\begin{cases} \dfrac{d\xi_1}{dt} = (-1)^{n-1} I_n \xi_n \\ \dfrac{d\xi_2}{dt} = \xi_1 - I_{n-1}\xi_n \\ \dfrac{d\xi_3}{dt} = \xi_2 + I_{n-2}\xi_n \\ \dots\dots\dots\dots\dots\dots\dots\dots \\ \dfrac{d\xi_n}{dt} = \xi_{n-1} + I_1\xi_n\,. \end{cases} \tag{7}$$

Then one has to introduce a symmetric quadratic form

$$V = \frac{1}{2}\sum_{i,j}^{n-1} A_{ij}\xi_i\xi_j \tag{8}$$

and evaluate its time derivative along the solution of (7), arriving to

$$\dot{V} = \sum A_{ij}\xi\dot{\xi}_j = \sum B_{ij}\xi_i\xi_j\,, \tag{9}$$

with well determined B_{ij}.

Step three

Require that the Hurwitz conditions guaranteeing that the eigenvalues of (7) have negative real parts, imply that either i) V is positive definite and $\dot{V} \leq 0$; or ii) V and $\dot{V}$ are positive definite when one of that conditions is not verified; or iii) when the Hurwitz conditions are not verified, V verifies theorem 3.

3 The case $n = 3$

For $n = 3$, (7) reduces to

$$\begin{cases} \dfrac{d\xi_1}{dt} = I_3\xi_3 \\ \dfrac{d\xi_2}{dt} = \xi_1 - I_2\xi_3 \\ \dfrac{d\xi_3}{dt} = \xi_2 + I_1\xi_3 . \end{cases} \tag{10}$$

We set

$$V = \frac{1}{2}(a\xi_1^2 + b\xi_2^2 + c\xi_3^2 + 2d\xi_1\xi_2 + 2e\xi_1\xi_3 + 2f\xi_2\xi_3) \tag{11}$$

with a, b, c, d, e, f real parameters to be determined. Because the characteristic equation is given by

$$\begin{vmatrix} -\lambda & 0 & I_3 \\ 1 & -\lambda & -I_2 \\ 0 & 1 & I_1 - \lambda \end{vmatrix} \equiv \lambda^3 - I_1\lambda^2 + I_2\lambda - I_3 = 0$$

the Hurwitz matrix is

$$\begin{pmatrix} -I_1 & -I_3 & 0 \\ 1 & I_2 & 0 \\ 0 & 0 & -I_3 \end{pmatrix}$$

and the Hurwitz's conditions for the stability {Cfr. [1], pag. 114}[1] are

$$I_1 < 0 \quad I_3 < 0 , \quad I_1 I_2 - I_3 < 0 , \tag{12}$$

one has to require that (12) imply the positive definiteness of V and the negative definiteness of $\frac{dV}{dt}$ (evaluated along the solution of (10)).

Since it turns out that

$$\begin{aligned} \frac{dV}{dt} &= a\xi_1\frac{d\xi_1}{dt} + b\xi_2\frac{d\xi_2}{dt} + c\xi_3\frac{d\xi_3}{dt} + \\ &+d(\frac{d\xi_1}{dt}\xi_2 + \xi_1\frac{d\xi_2}{dt}) + e(\frac{d\xi_1}{dt}\xi_3 + \xi_1\frac{d\xi_3}{dt} + f(\frac{d\xi_2}{dt}\xi_3 + \xi_2\frac{d\xi_3}{dt}) , \end{aligned} \tag{13}$$

one easily obtains

$$\frac{dV}{dt} = \begin{cases} d\xi_1^2 + f\xi_2^2 + (I_1c + I_3e - I_2f)\xi_3^2 + \\ (b+e)\xi_1\xi_2 + (I_3a - I_2d + I_1e + f)\xi_1\xi_3 \\ (I_1f - I_2b + I_3d + c)\xi_2\xi_3 . \end{cases} \tag{14}$$

[1]i.e. the conditions guaranteeing that the roots of the characteristic equation (eigenvalues of A) have al real negative parts

Requiring that

$$\begin{cases} d = -m_1 \\ f = -m_2 \\ I_1 c + I_3 e - I_2 f = -m_3 \\ b + e = 0 \\ I_3 a - I_2 d + I_1 e + f = 0 \\ I_1 f - I_2 b + I_3 d + c = 0\,, \end{cases} \tag{15}$$

it follows that

$$\frac{dV}{dt} = -\sum_{i=1}^{3} m_i \xi_i^2\,. \tag{16}$$

Our first aim is to choose m_i $(i = 1, 2, 3)$ in such a way that (12) imply

$$m_i \geq 0 \qquad (i = 1, 2, 3) \tag{17}$$

and look for solutions of (15) guaranteeing the positive definiteness of V. By virtue of (15), one obtains

$$\begin{cases} d = -m_1 \\ f = -m_2 \\ I_1 c + I_3 e = -m_3 - I_2 m_2 \\ b = -e \\ I_3 a + I_1 e = m_2 - m_1 I_2 \\ I_2 e + c = I_1 m_2 + I_3 m_1\,. \end{cases} \tag{18}$$

In view of $(18)_3$ and $(18)_6$ it follows that

$$\begin{cases} e = \dfrac{I_1(I_1 m_2 + I_3 m_1) + m_3 + I_2 m_2}{I_1 I_2 - I_3} < 0 \\ c = \dfrac{I_2(m_3 + I_2 m_2) + I_3(I_1 m_2 + I_3 m_1)}{I_3 - I_2 I_1} > 0 \\ b = -e > 0 \\ a = -\dfrac{I_1}{I_3} e + \dfrac{m_2}{I_3} - \dfrac{m_1 I_2}{I_3} \end{cases} \tag{19}$$

and hence one obtains

$$\begin{cases} a = -\dfrac{I_1}{I_3} e + \dfrac{(m_2 - m_1 I_2)}{I_3} \\ b = -e > 0 \\ c = \dfrac{I_2(m_3 + I_2 m_2) + I_3(I_1 m_2 + I_3 m_1)}{-(I_1 I_2 - I_3)} > 0 \\ d = -m_1 \leq 0 \\ e = \dfrac{I_1(I_1 m_2 + I_3 m_1) + m_3 + I_2 m_2}{I_1 I_2 - I_3} < 0 \\ f = -m_2 \leq 0\,. \end{cases} \tag{20}$$

In order to guarantee

$$a > 0 \tag{21}$$

one has to require

$$I_1 e + f - I_2 d > 0. \tag{22}$$

One can verify (22) in various manners. We now use the following choice:

$$I_1 e = -2f. \tag{23}$$

By virtue of (18), one obtains:

$$\begin{cases} a = \dfrac{I_2}{I_3} d + \dfrac{f}{I_3} \\[2ex] b = -e = \dfrac{I_1}{I_2} f + \dfrac{I_3}{I_2} d + \dfrac{c}{I_2} \\[2ex] c = -\dfrac{I_3}{I_1} e + \dfrac{I_2}{I_1} f - \dfrac{m_3}{I_1}. \end{cases} \tag{24}$$

Therefore choosing

$$m_3 = -2 I_1 I_3 f \tag{25}$$

one obtains

$$\begin{cases} a = \dfrac{I_2}{I_3} d + \dfrac{f}{2I_3} + \dfrac{f}{2I_3} \\[2ex] b = -e = \dfrac{I_1}{I_2} f + \dfrac{I_3}{I_2} d + \dfrac{c}{I_2} \\[2ex] c = -\dfrac{I_3}{I_1} e + \dfrac{I_2}{I_1} f + 2 I_3 f. \end{cases} \tag{26}$$

and hence

$$2V = \begin{cases} \dfrac{f}{2I_3}\xi_1^2 + \dfrac{2f}{I_1}\xi_2^2 - \dfrac{I_3}{I_1} e \xi_3^2 + \dfrac{d}{I_2 I_3}(I_2\xi_1 + I_3\xi_2)^2 \\[2ex] + \dfrac{f}{I_1 I_2}(I_1\xi_2 + I_2\xi_3)^2 + \dfrac{f}{2I_3}(2I_3\xi_1 + \xi_3)^2 \end{cases} \tag{27}$$

with

$$\begin{cases} f = -m_2, \quad d = -m_1 \\[2ex] e = -\dfrac{2f}{I_1}, \quad c = \left(2\dfrac{I_3}{I_1^2} + \dfrac{I_2}{I_1} + 2I_3\right) f \end{cases} \tag{28}$$

and

$$\dot{V} = -\left(m_1\xi_1^2 + m_2\xi_2^2 + 2I_1 I_3 m_1 \xi_3^2\right). \tag{29}$$

Therefore

$$2V = \begin{cases} \dfrac{f}{2I_3}\xi_1^2 + \dfrac{2f}{I_1}\xi_2^2 - \dfrac{2I_3}{I_1^2} f \xi_3^2 + \dfrac{d}{I_2 I_3}(I_2\xi_1 + I_3\xi_2)^2 \\[2ex] + \dfrac{f}{I_1 I_2}(I_1\xi_2 + I_2\xi_3)^2 + \dfrac{f}{2I_3}(2I_3\xi_1 + \xi_3)^2 \end{cases} \tag{30}$$

and for

$$d = f \tag{31}$$

$$2V = \begin{cases} \dfrac{f}{2I_1^2 I_2 I_3}[I_1^2 I_2 \xi_1^2 + 2I_1 I_2 I_3 \xi_2^2 - 2I_2 I_3 \xi_3^2 + \\ \\ +(I_2\xi_1 + I_3\xi_2)^2 + (I_1\xi_2 + I_2\xi_3)^2 + (2I_3\xi_1 + \xi_3)^2]. \end{cases} \tag{32}$$

Finally choosing

$$f = -4I_1^2 I_2 I_3 (I_1 I_2 - I_3)$$

one obtains

$$V = \begin{cases} I_1^2 I_2 \xi_1^2 + 2I_1 I_2 I_3 \xi_2^2 - 2I_2 I_3 \xi_3^2 + \\ \\ +(I_2\xi_1 + I_3\xi_2)^2 + (I_1\xi_2 + I_2\xi_3)^2 + (2I_3\xi_1 + \xi_3)^2 \end{cases} \tag{33}$$

$$\begin{aligned} \dot{V} &= d\xi_1^2 + f\xi_2^2 - m_3\xi_3^2 = f(\xi_1^2 + \xi_2^2 + 2I_1 I_3 \xi_3^2) = \\ &= 4I_1^2 I_2 I_3 (I_3 - I_1 I_2)(\xi_1^2 + \xi_2^2 + 2I_1 I_3 \xi_3^2). \end{aligned} \tag{34}$$

Theorem 4. *Let*

$$I_1 < 0\,, \quad I_3 < 0\,, \quad I_1 I_2 - I_3 < 0 \tag{35}$$

hold. Then:

i) V is positive definite,

ii) $\dot{V} < 0$ along the solutions.

Further if

$$I_1 > 0, \quad I_3 > 0, \quad I_1 I_2 - I_3 > 0 \tag{36}$$

then

1) V is no more positive definite but in any ball of the phase space centered at the origin exists a point in which $V > 0$;

2) $\dot{V} > 0$ along the solutions.

4 Another Liapunov functional for the case $n = 3$

Choosing

$$I_1 e = -3f \tag{37}$$

it turns out that

$$I_3 a + I_2 d + I_1 e + f = 0$$

is equivalent to

$$I_3 a + I_2 d + \frac{1}{2} I_1 e + \left(\frac{1}{2} I_1 e + f\right) = 0$$

i.e.

$$a = \frac{I_2}{I_3} d - \frac{1}{2}\frac{I_1}{I_3} e + \frac{f}{2I_3}. \tag{38}$$

On the other hand

$$c = -\frac{2I_3}{I_1}e + \frac{I_2}{I_1}f + \frac{1}{I_1}(I_3 e - m_3) \tag{39}$$

hence under the condition

$$m_3 > I_3 e = I_3\left(-\frac{3f}{I_1}\right) = -\frac{3I_3}{I_1}f = \frac{3I_3}{I_2}m_1 \tag{40}$$

one obtains

$$2V = \begin{cases} \frac{f}{2I_3}\xi_1^2 + \frac{3f}{I_1}\xi_2^2 + \frac{I_3 e - m_3}{I_1}\xi_3^2 + \frac{d}{I_2 I_3}(I_2\xi_1 + I_3\xi_2)^2 + \\ \\ +\frac{f}{I_1 I_2}(I_1\xi_2 + I_2\xi_3)^2 - \frac{e}{2I_1 I_3}(I_1\xi_1 - I_3\xi_3)^2 \end{cases} \tag{41}$$

with

$$\dot{V} = -(m_1\xi_1^2 + m_2\xi_2^2 + m_3\xi_3^2) \tag{42}$$

$$2V = \begin{cases} \frac{f}{2I_3}\xi_1^2 + \frac{3f}{I_1}\xi_2^2 + \frac{-3I_3 f - m_3 I_1}{I_1^2}\xi_3^2 + \frac{d}{I_2 I_3}(I_2\xi_1 + I_3\xi_2)^2 + \\ \\ +\frac{f}{I_1 I_2}(I_1\xi_2 + I_2\xi_3)^2 + \frac{3f}{2I_1^2 I_3}(I_1\xi_1 - I_3\xi_3)^2. \end{cases} \tag{43}$$

On choosing

$$m_3 = 3m_2 I_2 = -3fI_2 \tag{44}$$

it follows that

$$\frac{-3I_3 f - m_3 I_1}{I_1^2} = \frac{-3I_3 f + 3fI_1 I_2}{I_1^2} = \frac{3f(I_1 I_2 - I_3)}{I_1^2} \tag{45}$$

and hence for

$$d = f \tag{46}$$

$$2V = \begin{cases} \frac{f}{2I_1^2 I_2 I_3}[I_1^2\xi_1^2 + 6I_1 I_3\xi_2^2 + 6I_3(I_1 I_2 - I_3)\xi_3^2 + \\ \\ +(I_2\xi_1 + I_3\xi_2)^2 + (I_1\xi_2 + I_2\xi_3)^2 + 3(I_1\xi_1 - I_3\xi_3)^2]. \end{cases} \tag{47}$$

$$\dot{V} = -m_2(\xi_1^2 + \xi_2^2 + 3I_2\xi_3^2). \tag{48}$$

Choosing

$$f = 4I_1^2 I_3 I_2 \tag{49}$$

one obtains

$$\begin{aligned} V = {} & I_1^2\xi_1^2 + 6I_1 I_3\xi_2^2 + 6I_3(I_1 I_2 - I_3)\xi_3^2 + (I_2\xi_1 + I_3\xi_2)^2 + \\ & +(I_1\xi_2 + I_2\xi_3)^2 + 3(I_1\xi_1 - I_3\xi_3)^2. \end{aligned} \tag{50}$$

$$\dot{V} = 4I_1^2 I_2 I_3(\xi_1^2 + \xi_2^2 + 3I_2\xi_3^2)\,, \tag{51}$$

and the following theorem holds.

Theorem 5. *Let* (35) *holds. Then the function V given by* (50), *verifies i) and ii) of Theorem* 4. *Further if*

$$I_2 I_3 > 0\,, \tag{52}$$

1) V takes positive values in any ball containing the origin;

2) $\dot{V} > 0$ along the solutions.

5 Final Remarks

As has been done in the case $n = 2$ [4]-[6], the functionals (33), (50) will be generalized for the ternary reaction-diffusion systems of P.D.Es. [7].

Acknowledgements

This work has been performed under the auspicies of the G.N.F.M. of I.N.D.A.M. and M.I.U.R. (P.R.I.N. 2007): *Waves and Stability in Continuous Media.*

References

[1] David R. Merkin: "Introduction to the theory of stability". Text in Applied Mathematics **24**, Springer, 1997.

[2] J. La Salle, S. Lefschetz: "Stability by Liapunov's Direct Method". Academic Press, 1961.

[3] F. R. Gantmacher: "The theory of Matrices", vol. II, A.M.S. Chelsea Publishing, 2000.

[4] S. Rionero: "A rigorous reduction of the L^2-stability of the solutions to a nonlinear binary reaction - diffusion system of P.D.Es.". *J.Math.Anal.Appl.*, **319**, pp.377-397, 2006.

[5] S.Rionero: "L^2-stability of the solutions to a nonlinear binary reaction - diffusion system of P.D.Es.". *Rend. Mat. Acc. Lincei*, s. 9, **16**, pp.227-238, 2005.

[6] S.Rionero: "On instability sources in dynamical systems". *Trends and Applications of Mathematics to Mechanics (STAMM 2002).* S.Rionero, G. Romano (editors), Springer-Verlag Italia, pp.141-154, 2005.

[7] S.Rionero: "Reduction of the Liapunov Direct Method to the Liapunov first aproximation Method for reaction-diffusion systems of P.D.Es. (To appear).

A Classical Approach of a Multi-Temperature Mixture of Fluids

T. Ruggeri

Department of Mathematics and Research Center of Applied Mathematics (CIRAM), University of Bologna, Italy

Dedicato con tanto affetto a Brian Straughan per il suo 60 compleanno

Abstract

We present a classical approach of mixture of compressible fluids when each constituent has its own temperature. The introduction of an *average temperature* together with the entropy principle dictates the classical Fick law for diffusion and also new constitutive equations associated with the difference of temperatures between the components. The constitutive equations fit with results recently obtained through the *Maxwellian iteration* procedure in extended thermodynamics theory of multi-temperature mixtures. The differences of temperatures between the constituents imply the existence of a new *dynamical pressure* even if the fluids have a zero bulk viscosity.

1 Mixtures in rational thermodynamics

In the context of rational thermodynamics, the description of a homogeneous mixture of n constituents is based on the postulate that each constituent obeys to the same balance laws that a single fluid does [1]. The laws express the equations of balance of masses, momenta and energies

$$\left\{\begin{aligned} &\frac{\partial \rho_\alpha}{\partial t} + \operatorname{div}(\rho_\alpha \mathbf{v}_\alpha) = \tau_\alpha, \\ &\frac{\partial(\rho_\alpha \mathbf{v}_\alpha)}{\partial t} + \operatorname{div}(\rho_\alpha \mathbf{v}_\alpha \otimes \mathbf{v}_\alpha - \mathbf{t}_\alpha) = \mathbf{m}_\alpha, \qquad\qquad (\alpha = 1, 2, \ldots, n) \\ &\frac{\partial\left(\frac{1}{2}\rho_\alpha v_\alpha^2 + \rho_\alpha \varepsilon_\alpha\right)}{\partial t} + \operatorname{div}\left\{\left(\frac{1}{2}\rho_\alpha v_\alpha^2 + \rho_\alpha \varepsilon_\alpha\right)\mathbf{v}_\alpha - \mathbf{t}_\alpha \mathbf{v}_\alpha + \mathbf{q}_\alpha\right\} = e_\alpha. \end{aligned}\right. \tag{1}$$

On the left hand side, ρ_α is the density, $\mathbf{v}_\alpha$ is the velocity, ε_α is the internal energy, $\mathbf{q}_\alpha$ is the heat flux and $\mathbf{t}_\alpha$ is the stress tensor of the constituent α. The stress tensor $\mathbf{t}_\alpha$ can be decomposed into a pressure part $-p_\alpha \mathbf{I}$ and a viscous part $\mathbf{m}\sigma_\alpha$.

On the right hand sides τ_α, $\mathbf{m}_\alpha$ and e_α represent the production terms related to the interactions between constituents. Due to the total conservation of mass, momentum and energy of the mixture, the sum of production terms over all constituents must vanish

$$\sum_{\alpha=1}^{n} \tau_\alpha = 0, \quad \sum_{\alpha=1}^{n} \mathbf{m}_\alpha = \mathbf{0}, \quad \sum_{\alpha=1}^{n} e_\alpha = 0.$$

For the sake of simplicity, we ignore in the sequel the possibility of chemical reactions ($\tau_\alpha = 0$).

Mixture quantities $\rho, \mathbf{v}, \varepsilon, \mathbf{t}$ and $\mathbf{q}$ are defined ad hoc in such way when we take the summation of Eqs. (1), they are in the same form that the conservation laws of mass, momentum and energy of a single fluid.

In order to compare the balance equations of mixture and single fluid, we write Eqs. (1) in the equivalent form,

$$\begin{cases} \dfrac{\partial \rho}{\partial t} + \operatorname{div}(\rho \mathbf{v}) = 0, \\[2ex] \dfrac{\partial (\rho \mathbf{v})}{\partial t} + \operatorname{div}(\rho \mathbf{v} \otimes \mathbf{v} - \mathbf{t}) = \mathbf{0}, \\[2ex] \dfrac{\partial \left(\frac{1}{2}\rho v^2 + \rho\varepsilon\right)}{\partial t} + \operatorname{div}\left\{\left(\dfrac{1}{2}\rho v^2 + \rho\varepsilon\right)\mathbf{v} - \mathbf{t}\mathbf{v} + \mathbf{q}\right\} = 0, \\[2ex] \dfrac{\partial \rho_b}{\partial t} + \operatorname{div}(\rho_b \mathbf{v}_b) = \tau_b, \qquad\qquad (b = 1, \dots, n-1) \\[2ex] \dfrac{\partial (\rho_b \mathbf{v}_b)}{\partial t} + \operatorname{div}(\rho_b \mathbf{v}_b \otimes \mathbf{v}_b - \mathbf{t}_b) = \mathbf{m}_b, \\[2ex] \dfrac{\partial \left(\frac{1}{2}\rho_b v_b^2 + \rho_b\varepsilon_b\right)}{\partial t} + \operatorname{div}\left\{\left(\dfrac{1}{2}\rho_b v_b^2 + \rho_b\varepsilon_b\right)\mathbf{v}_b - \mathbf{t}_b\mathbf{v}_b + \mathbf{q}_b\right\} = e_b, \end{cases} \tag{2}$$

where the index b runs from 0 to $n-1$ and the first 5 euqations are the conservation of total mass, momentum and energy.

In this multi-temperature model (MT), used in particular in plasma physics [2], we have $5n$ independent field variables ρ_α, $\mathbf{v}_\alpha$ and T_α ($\alpha = 1, 2, \dots, n$), where T_α is the temperature of constituent α. To close the system (2) of the field equations of the mixture process, we must write the constitutive equations for the quantities $p_\alpha, \varepsilon_\alpha, \mathbf{q}_\alpha, \sigma_\alpha$ ($\alpha = 1, 2, \dots, n$) and $\mathbf{m}_b$, e_b ($b = 1, \dots, n-1$) in terms of the field variables $\rho_\alpha, \mathbf{v}_\alpha$ and T_α ($\alpha = 1, 2, \dots, n$).

Due to the difficulties to measure the temperature of each component, a common practice among engineers and physicists is to consider only one temperature for the mixture. When we use a single temperature (ST), Eq. $(2)_6$ disappears and we get a unique global conservation of the total energy in the form $(2)_3$ (see for example [3]). In a recent paper, Ruggeri and Simić [4] discussed the mathematical difference between the ST and the MT models when the fluid components are Eulerian gases ($\mathbf{q}_\alpha = 0, \sigma_\alpha = 0$).

They proved that the differential system of the ST model is a *principal sub-system* [5] of the MT model, and for large time, MT solutions converge to ST ones.

A further step of coarsening theory is the classical approach of mixtures, in which the independent field variables are the density, the mixture velocity, the individual temperature of the mixture and the concentrations of constituents. In that case system also $(2)_5$ is taken out.

To reveal the relation between extended and classical models, a formal iterative scheme known as *Maxwellian iteration* is applied (see e.g [3]). In the case of ST model the first iterates $\mathbf{J}_a^{(1)}$ are calculated from the right–hand sides of balance laws $(2)_5$ by putting "zeroth" iterates — equilibrium values $\mathbf{J}_a^{(0)} = \mathbf{0}$ on the left–hand sides. The next step second iterates $\mathbf{J}_a^{(2)}$ is obtained from the right–hand sides of the same Eqs. by putting first iterate $\mathbf{J}_a^{(1)}$ on their left–hand sides, an so on. If we apply the first *Maxwellian iteration* the Fick laws of diffusion fluxes are obtained. Roughly speaking, the Fick laws are obtained by neglecting the accelerations of the relative motions of the constituents and the classical theory is an approximation of the ST model (see [3], Chapter 5).

In a recent paper, Ruggeri and Simić [6] considered the Maxwellian iteration of system (2) in the case of a binary mixture of Eulerian fluids; they obtained the Fick laws as a first order term of the expansion of the component momentum equations. When each component has its own temperature, an additive constitutive equation comes from a limit case of the constituent equation of energy in the form

$$\Theta = L_\theta\,(\gamma_1 - \gamma_2)\,\mathrm{div}\,\mathbf{v} \tag{3}$$

where L_θ is a new phenomenological positive coefficient and

$$\Theta = T_2 - T_1\,.$$

2 A Variant of Classical approach for multi-temperature mixture

Eq. (3) is not obtained in classical theory. It is necessary to find a variant form of classical approach to recover also equations like (3) in the general case of mixtures with n non necessarily Eulerian compressible fluids.

In the classical approach, the velocity field $\mathbf{v}$ corresponds to an average velocity with respect to mass components. Thanks to Fick Laws, the diffusion fluxes $\mathbf{J}_\alpha = \rho_\alpha\,(\mathbf{v}_\alpha - \mathbf{v})$ determine the component velocities $\mathbf{v}_\alpha$.

By analogy with the velocity fields a natural extension of the classical approach is to consider an average temperature T and $\Theta_\alpha = T_\alpha - T$ $(\alpha = 1, \cdots, n)$ as constitutive quantities; similarly with the diffusion velocity fluxes, we name Θ_α *the diffusion temperature fluxes.*

To define an average temperature T, Ruggeri and Simić assumed the total intrinsic internal energy of the mixture (that coincides with the full internal energy for processes not so far from equilibrium) to be the same in the multi-temperature case than in the ST model when the temperature is T [7]. Consequently, T is defined through the local implicit solution of the equation

$$\rho\varepsilon \equiv \sum_{\alpha=1}^{n} \rho_\alpha \varepsilon_\alpha(\rho_\alpha, T) = \sum_{\alpha=1}^{n} \rho_\alpha \varepsilon_\alpha(\rho_\alpha, T_\alpha). \tag{4}$$

Such a choice comes from the case of particular classes of solutions for perfect gases [7] and from the fact that the equation of energy governs the evolution of the common temperature T for the ST model. The consequences on the entropy of the mixture will confirm the physical ground of Eq. (4).

We consider as in usual classical approach the field equations: conservation of total, mass, momentum and energy and balance law of mass for each constituent (we neglect chemical reaction for the moment)

$$\begin{cases} \dfrac{d\rho}{dt} + \rho \operatorname{div} \mathbf{v} = 0, \\ \rho \dfrac{d\mathbf{v}}{dt} - \operatorname{div} \mathbf{t} = 0, \\ \rho \dfrac{d\varepsilon}{dt} - \mathbf{t} \ \operatorname{grad} \mathbf{v} + \operatorname{div} \mathbf{q} = 0, \\ \rho \dfrac{dc_b}{dt} + \operatorname{div} \mathbf{J}_\mathrm{b} = 0, \qquad (\mathrm{b} = 1, \cdots \mathrm{n} - 1), \end{cases} \tag{5}$$

where

$$\frac{d}{dt} = \frac{\partial}{\partial t} + \mathbf{v} \cdot \frac{\partial}{\partial \mathbf{x}}$$

represents the material derivative of the mixture motion

$$c_\alpha = \frac{\rho_\alpha}{\rho}, \qquad \left(\sum_{\alpha=1}^{n} c_\alpha = 1 \right)$$

are the components concentrations, and

$$\mathbf{J}_\alpha = \rho_\alpha \mathbf{u}_\alpha = \rho_\alpha (\mathbf{v}_\alpha - \mathbf{v}), \qquad \left(\sum_{\alpha=1}^{n} \mathbf{J}_\alpha = 0 \right)$$

are the diffusion fluxes of the components.

Taking into account Eqs. $(5)_1$, $(5)_4$ and (4), Eq. $(5)_3$ of energy of the mixture can be written as a differential equation for the average temperature T

$$\rho \varepsilon_{,T} \frac{dT}{dt} = \rho^2 \varepsilon_{,\rho} \operatorname{div} \mathbf{v} + \sum_{\mathrm{b}=1}^{\mathrm{n}-1} \varepsilon_{,c_\mathrm{b}} \operatorname{div} \mathbf{J}_\mathrm{b} - \mathbf{t} \ \operatorname{grad} \mathbf{v} - \operatorname{div} \mathbf{q} \tag{6}$$

where the comma denotes the partial derivative with respect to the subscript.
As $\mathbf{J}_\alpha$ $(\alpha = 1, \cdots, n)$ are associated with the difference between component and average velocities

$$\Theta_\alpha = T_\alpha - T \quad (\alpha = 1, \cdots, n)$$

corresponding to the difference between component and average temperatures are non-equilibrium thermodynamical variables. Near the equilibrium, Eq. (4) can be expanded to the first order; then

$$\sum_{\alpha=1}^{n} \rho_\alpha c_V^{(\alpha)} \Theta_\alpha = 0 \quad \Longleftrightarrow \quad T = \frac{\sum_{\alpha=1}^{n} \rho_\alpha c_V^{(\alpha)} T_\alpha}{\sum_{\alpha=1}^{n} \rho_\alpha c_V^{(\alpha)}} \tag{7}$$

where

$$c_V^{(\alpha)} = \frac{\partial \varepsilon_\alpha}{\partial T_\alpha} (\rho_\alpha, T)$$

denotes the specific heat at constant volume for the constituent α at equilibrium. Consequently Eq. (7) yields

$$\Theta_n = -\frac{1}{\rho_n c_V^{(n)}} \sum_{b=1}^{n-1} \rho_b \, c_V^{(b)} \Theta_b. \tag{8}$$

The definition of the total specific entropy S of the mixture is

$$\rho S = \sum_{\alpha=1}^{n} \rho_\alpha S_\alpha(\rho_\alpha, T_\alpha) \tag{9}$$

where S_α $(\alpha = 1, \cdots, n)$ are the specific entropies of the components. It is possible to prove taking the Gibbs relation that at the first order expansion the specific entropy S depends only on T and not on Θ_b $(b = 1, \ldots, n-1)$.
On the contrary, a first order expansion of the total pressure of the mixture $p = \sum_{\alpha=1}^{n} p_\alpha(\rho_\alpha, T_\alpha)$ together with Eq. (8) yields

$$p = p_0 + \pi_\theta, \tag{10}$$

where

$$p_0 = \sum_{\alpha=1}^{n} p_\alpha(\rho_\alpha, T), \quad \pi_\theta = \sum_{b=1}^{n-1} r_b \, \Theta_b$$

and

$$r_b = \frac{1}{\rho_n c_V^{(n)}} \left\{ \rho_n c_V^{(n)} \frac{\partial p_b}{\partial T_b} (\rho_b, T) - \rho_b c_V^{(b)} \frac{\partial p_n}{\partial T_n} (\rho_n, T) \right\} \quad (b = 1, \ldots, n-1).$$

Therefore, the total pressure p of the mixture is a sum of the equilibrium part p_0 depending on T and a new dynamical pressure part (as a non-equilibrium term) π_θ due to the difference of temperatures between the constituents.

We emphasis the fact that the entropy density depends only on T and not on the Θ_a justifies - in the present theory - the Θ_a have to be considered as constitutive quantities (undetermined quantities). This is a key point in the model.
Consequently, the aim of the next section consists in determining the constitutive equations for the Θ_a by using the entropy principle and *TIP* arguments.

3 Entropy principle and constitutive equations

We assume the internal energy $\varepsilon(\rho, T, c_b)$ and the equilibrium pressure $p_0(\rho, T, c_b)$ satisfy the Gibbs equation

$$TdS = d\varepsilon - \frac{p_0}{\rho^2} d\rho - \sum_{b=1}^{n-1} (\mu_b - \mu_n)\ dc_b \,. \tag{11}$$

where the component chemical potentials in equilibrium are

$$\mu_\alpha = \varepsilon_\alpha(\rho_\alpha, T) + \frac{p_\alpha(\rho_\alpha, T)}{\rho_\alpha} - TS_\alpha(\rho_\alpha, T).$$

As in the classical case with a single temperature, the time derivative can be eliminated between Eqs $(5)_1$, $(5)_4$ and (6) and we obtain the balance law of entropy:

$$\rho\frac{dS}{dt} + \operatorname{div}\left\{\frac{1}{\mathrm{T}}\left(\mathbf{q} - \sum_{\mathrm{b}=1}^{\mathrm{n}-1}(\mu_{\mathrm{b}} - \mu_{\mathrm{n}})\,\mathbf{J}_{\mathrm{b}}\right)\right\} =$$

$$\mathbf{q}\cdot\operatorname{grad}\left(\frac{1}{\mathrm{T}}\right) - \sum_{\mathrm{b}=1}^{\mathrm{n}-1}\mathbf{J}_{\mathrm{b}}\cdot\operatorname{grad}\left(\frac{\mu_{\mathrm{b}} - \mu_{\mathrm{n}}}{\mathrm{T}}\right) + \frac{1}{\mathrm{T}}\operatorname{tr}\left(\mathbf{t}\,\mathbf{D}^{\mathrm{D}}\right) - \frac{1}{\mathrm{T}}\sum_{\mathrm{b}=1}^{\mathrm{n}-1}\Theta_{\mathrm{b}}\,\mathrm{r}_{\mathrm{b}}\operatorname{div}\mathbf{v}$$

We observe that the entropy production is the sum of products of the following quantities:

thermodynamic fluxes	**thermodynamic forces**
heat flux $\mathbf{q}$	temperature gradient grad $\left(\frac{1}{\mathrm{T}}\right)$,
diffusion fluxes $\mathbf{J}_b$	chemical potential gradients grad $\left(\frac{\mu_{\mathrm{b}}-\mu_{\mathrm{n}}}{\mathrm{T}}\right)$,
stress deviator $\mathbf{t}^D$	deviatoric velocity gradient $\mathbf{D}^D$,
diffusive temperature fluxes Θ_b	r_b times divergence of velocity $r_b \operatorname{div}\mathbf{v}$

In accordance with the case of a single temperature model [8, 9], [3] chapter 5, in *TIP* near equilibrium, the fluxes depend linearly on the associated forces. For the stress deviator, the heat flux and the diffusion fluxes, we obtain the classical constitutive equations of Navier-Stokes, Fourier and Fick; moreover, for the diffusive temperature fluxes we get

$$\Theta_a = -\sum_{b=1}^{n-1} M_{ab}\, r_b \operatorname{div}\mathbf{v} \qquad (\mathrm{a} = 1, \cdots, \mathrm{n} - 1). \tag{12}$$

The production of entropy must be non-negative [10, 11]; consequently, the *phenomenological coefficients* must satisfy the usual inequalities and additively assuming the Onsager symmetry property, $M_{ab} = M_{ba}$ $(a, b = 1, \cdots, n - 1)$, we deduce that coefficients M_{ab} are *associated with a positive definite quadratic form.*

Therefore the results are the same as in classical theory but we get additively the new constitutive equations (12) for the difference of the temperatures.

Substituting Eq. (12) into the Eq. (10) of pressure, we obtain

$$p = p_0 + \pi_\theta$$

with a dynamical pressure π_θ such that

$$\pi_\theta = -L_\pi \,\mathrm{div}\,\mathbf{v} \qquad \text{with} \qquad \mathrm{L}_\pi = \sum_{\mathrm{a,b=1}}^{\mathrm{n-1}} \mathrm{M_{ab}}\ \mathrm{r_a r_b} \geq 0.$$

We have considered the simple case of Stokes fluids. If the fluid is non Stokesian, the Navier-Stokes stress tensor of viscosity is

$$\sigma = \lambda\ (\mathrm{div}\,\mathbf{v})\,\mathbf{I} + 2\,\nu\ \mathbf{D}^{\mathrm{D}},$$

where λ is the bulk viscosity. The stress tensor $\mathbf{t}$ becomes

$$\mathbf{t} = -(p_0 + \pi_\theta)\,\mathbf{I} + \sigma = -p\ \mathbf{I} + 2\,\nu\ \mathbf{D}^D,$$

with

$$p = p_0 + \pi_\theta + \pi_\sigma.$$

The non-equilibrium pressure $p - p_0$ is separated in two different parts. The first one $\pi_\sigma = -\lambda\ \mathrm{div}\,\mathbf{v}$ is related to the bulk viscosity and the second one $\pi_\theta = -L_\pi\,\mathrm{div}\,\mathbf{v}$ is related to the multi-temperature effects between components.

4 Conclusion

In the case of fluid mixtures with different temperatures, a classical framework allows to determine the new constitutive equation (12) for the difference of temperatures in addition to the classical Navier-Stokes, Fourier and Fick laws. The difference of the temperatures produces a new dynamical pressure term which has a physical meaning and consequently can be measured. We point out that the dynamical pressure associated with multi-temperature fluid mixtures exists even if the bulk viscosity is null. For rarefied gases, it is well known that the bulk viscosity λ is null (Stokes fluids) and consequently no dynamical pressure exists in a classical mixture with one temperature; in a relativistic context, this pressure exists but remains very small [3]. Due to a non-zero dynamical pressure even for Stokes fluids, we conclude that multi-temperature mixtures of fluids have a great importance.

We focus attention on the fact that the present results are obtained thanks to an average temperature corresponding to the same internal energy than for the single temperature model.
It is important to note that the total energy conservation law yields the evolution of T, the entropy principle can be used due to the fact the entropy density S depends only on T and the Gibbs equation is always defined by using Eq. (11).

The phenomenological coefficients can me measured using kinetic considerations. More details of these results can be appear in ([12]).

Acknowledgment

This paper was supported by fondi MIUR Progetto di interesse Nazionale *Problemi Matematici Non Lineari di Propagazione e Stabilità nei Modelli del Continuo* Coordinatore T. Ruggeri, by the GNFM-INdAM.

References

[1] C. Truesdell, *Rational Thermodynamics* (McGraw-Hill, New York, 1969).

[2] T.K. Bose, *High Temperature Gas Dynamics* (Springer, Berlin, 2003).

[3] I. Müller and T. Ruggeri, *Rational Extended Thermodynamics* (Springer, New York, 1998).

[4] T. Ruggeri and S. Simić, Math. Meth. Appl. Sci. **30**, 827 (2007).

[5] G. Boillat and T. Ruggeri, Arch. Rational Mech. Anal. **137**, 305 (1997).

[6] T. Ruggeri and S. Simić, in *Asymptotic Methods in Non Linear Wave Phenomena*, edited by T. Ruggeri and M. Sammartino (World Scientific, Singapore, 2007), p. 186.

[7] T. Ruggeri and S. Simić, in *Proceedings Mathematical Physics Models and Engineering Sciences* (Liguori Editore, Napoli, 2008), p. 455.

[8] I. Müller, Z. Naturforsch. **28**a, 1801 (1973).

[9] I. Müller, *Thermodynamics* (Pitman, New York, 1985).

[10] B. D. Coleman and W. Noll, Arch. Rational Mech. Anal. **13**, 167 (1963).

[11] I. Müller, Arch. Rational Mech. Anal. **26**, 118 (1967).

[12] H. Gouin, T. Ruggeri, *Identification of an average temperature and a dynamical pressure in a multi-temperature mixture of fluids.* Phys. Rev. E (in press 2008).

On the Displacement Problem of Plane Linear Elastostatics

*R. Russo**

Dipartimento di Matematica,
Seconda Università di Napoli,
via Vivaldi 43, 81100 Caserta, Italy.

Abstract

We consider the displacement problem of linear elastostatics in a Lipschitz exterior domain of $\mathbb{R}^2$. We prove that if the boundary datum a lies in $L^2(\partial\Omega)$, then the problem has a unique very weak solution which converges to an assigned constant vector u_∞ at infinity if and if a and u_∞ satisfy a suitable compatibility condition.

Keywords: system of linear elastostatics, displacement problem, exterior Lipschitz domains.

1 Introduction

Consider the displacement (or Dirichlet) problem of classical elastostatics [6]

$$\begin{aligned} \Delta u + \frac{1}{1-2\nu}\nabla(\operatorname{div} u) = 0 \quad &\text{in } \Omega, \\ u = a \quad &\text{on } \partial\Omega, \\ \lim_{r\to+\infty} u(x) = u_\infty & \end{aligned} \tag{1}$$

in an exterior domain Ω of $\mathbb{R}^2$ defined by

$$\Omega = \mathbb{R}^2 \setminus \Omega' \tag{2}$$

with $\partial\Omega$ Lipschitz closed and connected curve, $u : \Omega \to \mathbb{R}^2$ is the (unknown) displacement field, u_∞ is an assigned nonzero constant vector, ν the Poisson ratio and $a \in L^2(\partial\Omega)$ is an assigned field on $\partial\Omega$. It is well–known that, in general, (1) does not admits a solution. To see quickly this, let us first observe that a solution to equations $(1)_1$ is biharmonic in a neighborhood of infinity and recall that by a results of M. Picone [10], if u converges to a constant vector u_0 at infinity, then

$$u_0 = \frac{1}{2\pi}\int_0^{2\pi} u(R,\theta). \tag{3}$$

*Email: remigio.russo@unina2.it

Therefore, if $\partial\Omega$ is a disk C_R, then (4) implies that in order to (4) be solvable a and u_∞ must satisfy the necessary *compatibility condition*

$$u_\infty = \frac{1}{2\pi}\int_0^{2\pi} a(R,\theta). \tag{4}$$

In particular, if $a = 0$, then problem (1) does not admits a solution. Therefore, a natural question arises[1]:

- *How do we must assign a and u_∞ in order that problem* (1) *has a solution in an arbitrary Lipschitz exterior domain of the plane?.*

In this note we study the above problem in the more general context of the displacement problem of homogeneous anisotropic linear elastostatics:

$$\begin{aligned} \operatorname{div}\mathbb{C}[\nabla u] &= 0 \quad \text{in } \Omega, \\ u &= a \quad \text{on } \partial\Omega, \\ \lim_{r\to+\infty} u(x) &= u_\infty \end{aligned} \tag{5}$$

in an exterior domain of $\mathbb{R}^2$, where $\mathbb{C}$ is the elasticity tensor. We prove that if $\mathbb{C}$ is positive definite and $a \in L^2(\partial\Omega)$, then system (5) has a unique *very weak solution* if and only if

$$\int_{\partial\Omega} (u_\infty - a)\cdot\psi = 0 \quad \forall\psi \in \mathfrak{C}. \tag{6}$$

where $\mathfrak{C}$ is the two dimensional linear space of the densities of the elastic simple layer potentials whose trace on $\partial\Omega$ are constant.

2 The elastic simple layer potential

In (5) $\mathbb{C}$ is a linear map from the space of the second–order tensors Sym into itself we assume to be *symmetric* , *i.e.*,

$$A\cdot\mathbb{C}[B] = B\cdot\mathbb{C}[A], \quad \forall\, A,B \in \text{Sym} \;\Leftrightarrow\; C_{ijhk} = C_{hkij} \tag{7}$$

and *positive definite*, *i.e*,

$$A\cdot\mathbb{C}[A] = C_{ijhk}A_{ij}A_{hk} \geq \mu_m\,|A|^2, \quad \forall\, A \in \text{Sym}, \tag{8}$$

It is well–known that, under hypotheses (7), (8) the system

$$\operatorname{div}\mathbb{C}[\nabla u] = 0 \Leftrightarrow \partial_j(C_{ijhk}\partial_k u_h) = 0 \tag{9}$$

admits a fundamental solution $\mathfrak{Q}(x-y)$ which can be written

$$\mathfrak{Q}(z) = \Phi_0 \log|z| + \Phi'(z), \tag{10}$$

[1]In [5] G.P. Galdi and C. Simader deal with the same problem for the Stokes equations in a variational setting for domains of class C^2. Their technique can be also applied to system (5) in regular domains.

with $\Phi_0 \in \mathrm{Sym}$ and Φ' homogeneous symmetric second–order tensor field of order zero. The field $\mathcal{Q}(x-y)$ satisfies system (9) in $\mathbb{R}^2$, for every $x \neq y$. A standard argument assures that if $u \in W^{1,2}_{\mathrm{loc}}(\overline{\Omega})$ is a solution to system (9), such that $u = o(r)$, then there is a constant vector u_0 such that

$$u = o(r) \;\Rightarrow\; u(x) = u_0 + \mathcal{Q}(x)\sigma + \omega(x) \tag{11}$$

for large $|x|$, where

$$\sigma_i = \int_{\partial\Omega} e_i \cdot \mathbb{C}[\nabla u]n, \;\; i = 1,2; \quad \nabla_m \omega = O(r^{-1-m}), \tag{12}$$

and n is exterior (with respect to Ω') unit normal to $\partial\Omega$. Note that from (11) it easily follows that if $\nabla u \in L^2(\Omega)$ and u vanishes on $\partial\Omega$ then $u \equiv 0$. We say that a solution $u \in C^\infty(\Omega)$ to system (9) is *variational* provided $u \in W^{1,2}_{\mathrm{loc}}(\overline{\Omega})$. The following result is proved in [9], Ch. 5.

Lemma 1. *Let Ω_0 be a bounded domain of $\mathbb{R}^2$ and let $\mathbb{C}$ be symmetric and positive positive. If u is a variational solution to system* (9) *and* $\mathrm{tr}u_{|\partial\Omega_0} \in W^{1,2}(\partial\Omega_0)$, *then* $\mathbb{C}[\nabla u]n \in L^2(\partial\Omega_0)$.

Function (10) allows us to define the simple layer potential

$$v[\psi](x) = \int_{\partial\Omega} \mathcal{Q}(x-\zeta)\psi(\zeta)\mathrm{d}v_y.$$

with density $\psi \in L^2(\partial\Omega)$. Well–known results about singular integrals (see, *e.g.* [1], [4]) imply that $v[\psi]$ converges nontangentially[2]on *both* sided of $\partial\Omega$ to the same limit $\mathcal{S}[\psi]$ and $\mathcal{S}$ maps boundedly $L^2(\partial\Omega)$ into $W^{1,2}(\partial\Omega)$; the adjoint operator $\mathcal{S}^*$ is the unique continuos extension of $\mathcal{S}$ from $W^{-1,2}(\partial\Omega)$ into $L^2(\partial\Omega)$. Moreover $v[\psi] \xrightarrow{\mathrm{nt}} \mathcal{S}^*[\psi]$, for every $\psi \in W^{-1,2}(\partial\Omega)$, where now

$$v[\psi](x) = \langle \psi, \mathcal{Q}(x-\zeta)\rangle.$$

The trace of the traction of a simple layer potential with density $\psi \in W^{-1,2}(\partial\Omega)$, is linear and continuous from $W^{-1,2}(\partial\Omega)$ into itself and, denoting by $\mathbb{C}[\nabla v[\psi]]n^+$, $\mathbb{C}[\nabla v[\psi]]n^-$ its values on the *interior* and *exterior* faces of $\partial\Omega$, it holds

$$\psi = \mathbb{C}[\nabla v[\psi]]n^+ - \mathbb{C}[\nabla v[\psi]]n^-. \tag{13}$$

Of course, by interpolation $\mathcal{S}$ extends uniquely to a continuous operator from $W^{-\alpha,2}(\partial\Omega)$ into $W^{-\alpha+1,2}(\partial\Omega)$ for $\alpha \in [0,1]$. Moreover [3],

$$\begin{aligned} \psi \in W^{-1,2}(\partial\Omega) \;&\Rightarrow\; v[\psi] \in W^{1/2,2}_{\mathrm{loc}}(\overline{\Omega}) \\ \psi \in L^2(\partial\Omega) \;&\Rightarrow\; v[\psi] \in W^{3/2,2}_{\mathrm{loc}}(\overline{\Omega}). \end{aligned}$$

It holds [11]

[2]Let $\{\tau(\xi)\}_{\xi\in\partial\Omega}$ be a family of circular finite (not empty) triangles with vertex at ξ such that one component (except ξ) is contained in Ω. Let χ be a function in Ω. We say that $\chi(x)$ converges nontangentially at the boundary if

$$\chi(\xi) = \lim_{\substack{x\to\xi \\ (x\in\gamma(\xi))}} \chi(x) \Leftrightarrow \chi(x) \xrightarrow{\mathrm{nt}} \chi(\xi),$$

for almost all $\xi \in \partial\Omega$

Lemma 2. *Let Ω be a Lipschitz exterior domain of $\mathbb{R}^2$ and let $\mathbb{C}$ be constant, symmetric and positive definite. If $\mathcal{S}[\psi] = \text{constant}$, with $\psi \in L^2(\Omega)$ and $\int_{\partial\Omega} \psi = 0$, then $v[\psi] = 0$ in $\mathbb{R}^2$.*

Our results are based on the following lemma whose proof follows the argument lines in [11] (see also [12]).

Lemma 3. *Let Ω be a Lipschitz exterior domain of $\mathbb{R}^2$ and let $\mathbb{C}$ be constant, symmetric and positive definite. Then $\mathcal{S}$ is Fredholm with index zero and*

$$\dim \operatorname{Kern} \mathcal{S} = \gamma \leq 2. \tag{14}$$

Proof. Consider first the operator

$$\mathcal{S}_0 : W^{-1/2,2}(\partial\Omega) \to W^{1/2,2}(\partial\Omega)$$

In virtue of (13), the trace theorem and well–known estimates we have

$$\begin{aligned}
\|\psi\|_{W^{-1/2,2}(\partial\Omega)} &\leq \|\mathbb{C}[\nabla v[\psi]]n^+\|_{W^{-1/2,2}(\partial\Omega)} + \|\mathbb{C}[\nabla v[\psi]]n^-\|_{W^{-1/2,2}(\partial\Omega)} \\
&\leq c\left\{\|\nabla v[\psi]\|_{L^2(\Omega)} + \|\nabla v[\psi]\|_{L^2(\Omega')} + \|\nabla v[\psi]\|_{L^2(\complement\Omega\cap S_R)}\right\} \\
&\leq c\left\{\|\mathcal{S}[\psi]\|_{W^{1/2,2}(\partial\Omega)} + \|\mathcal{S}[\psi]\|_{W^{1/2,2}(\partial S_R)}\right\} \\
&\leq c\left\{\|\mathcal{S}[\psi]\|_{W^{1/2,2}(\partial\Omega)} + \|\mathcal{C}\|\right\}
\end{aligned}$$

where $\mathcal{C}$ is a compact operator from $W^{-1/2,2}(\partial\Omega)$ into a Banach space. Hence, by a classical theorem of J. Peetre (see, *e.g*, [8]), taking into account that $\mathcal{S}_0$ is self–adjoint, it follows that $\mathcal{S}_0$ is Fredholm with index zero. Let $\psi(\neq 0) \in \operatorname{Kern} \mathcal{S}_0$. It is obvious that $v[\psi] = 0$ in Ω'. Let $\{\psi_i\}_{i=1,2,3} \subset \operatorname{Kern} \mathcal{S}_0$, with $\psi_i \neq 0$, then necessarily there are three scalars α_i such that $\alpha_i(\psi_i)_{\partial\Omega} = 0$. Then, $\psi = \alpha_i\psi_i$, $v[\psi]_{|\partial\Omega} = 0$ and $v[\psi] = O(r^{-1})$ so that by uniqueness $v[\psi] = 0$ in $\mathbb{R}^2$ so that by (13) $\psi = 0$. Hence, it follows that the system $\{\psi_i\}_{i=1,2,3}$ is linearly dependent and (14) is proved taking into account that by Lemma 1 $\operatorname{Kern} \mathcal{S}_0 \in W^{1,2}(\partial\Omega)$. Let $a \in W^{1,2}(\partial\Omega)$ satisfy

$$\int_{\partial\Omega} a \cdot \psi = 0, \quad \forall \psi \in \operatorname{Kern} \mathcal{S}_0.$$

By Lemma 3 , there is a density $\psi \in W^{-1/2,2}(\partial\Omega)$ such that $\mathcal{S}_0[\psi] = a$. In virtue of Lemma 1 $\mathbb{C}[\nabla v[\psi]]n^\pm \in L^2(\partial\Omega)$ so that by (13) $\psi \in L^2(\partial\Omega)$. Therefore the operator $\mathcal{S}$ has a closed range and $\operatorname{Kern} \mathcal{S} = \operatorname{Kern} \mathcal{S}_0$. If $\psi \in \operatorname{Kern} \mathcal{S}^*$, denote by $\{\psi_k\}_{k\in\mathbb{N}}$ a sequence in $L^2(\partial\Omega)$ which converges strongly in $L^2(\partial\Omega)$ to ψ. It is not hard to see that

$$\int_{\Omega'} \varphi \cdot v[\psi_k] = -\int_{\partial\Omega} \mathcal{S}[\psi_k] \cdot \mathbb{C}[\nabla z]n \tag{15}$$

for all $\varphi \in C_0^\infty(\Omega)$, with z solution to the system[3]

$$\begin{aligned}
\operatorname{div} \mathbb{C}[\nabla z] + \varphi = 0 \quad &\text{in } \Omega', \\
z = 0 \quad &\text{on } \partial\Omega.
\end{aligned} \tag{16}$$

[3]Existence of a solution to problem (16) is achieved by the classical argument to reduce (16) to a boundary value problem by the volume potential on Ω with density φ.

Then, letting $k \to +\infty$ in (15), we get $v[\psi] = 0$ in Ω'. Moreover, let z be a the solution to the system

$$\begin{aligned} \operatorname{div} \mathbb{C}[\nabla z] + \varphi = 0 \quad & \text{in } \Omega, \\ z = 0 \quad & \text{on } \partial\Omega. \end{aligned} \tag{17}$$

with $\varphi \in C_0^\infty(\Omega)$ such that

$$\int_\Omega \varphi \cdot v[\psi'] = 0, \quad \psi' \in \operatorname{Kern} \mathcal{S}_0. \tag{18}$$

Note that (18) implies that $z = o(1)$ at infinity. By a simple integration by parts on Ω_R we get

$$\begin{aligned} \int_\Omega \varphi \cdot v[\psi_k] = & \int_{\partial\Omega} a_k \cdot \mathbb{C}[\nabla z] n - \int_{\partial\Omega} z \cdot T\big[u_k\big] n \\ & + \int_{\partial C_R} u_k \cdot \mathbb{C}[\nabla z] e_R - \int_{\partial C_R} z \cdot \mathbb{C}[\nabla u] e_R \end{aligned}$$

Hence, taking into account the behavior at infinity of u_k, z and their derivatives, it follows

$$\int_\Omega \varphi \cdot v[\psi_k] = \int_{\partial\Omega} \mathcal{S}[\psi_k] \cdot \mathbb{C}[\nabla z] n \tag{19}$$

Letting $k \to +\infty$, in (19) yields

$$\int_\Omega \varphi \cdot v[\psi] = 0$$

for all $\varphi \in C_0^\infty(\Omega)$ which satisfies (18). Then necessarily $\psi \in \operatorname{Kern} \mathcal{S}_0$ and the Lemma is proved.

Lemma 4. *The linear spaces*

$$\mathfrak{C} = \{\psi' \in L^2(\partial\Omega) : \mathcal{S}[\psi'] = \text{constant}\}.$$

have dimension two.

Proof. If $\psi'(\neq 0) \in \mathfrak{C}$, then $\psi'_{\partial\Omega} \neq 0$; otherwise by Lemma 2 and (13) we have $\psi = 0$. Let $\{\psi'_i\}_{i=1,2,3} \subset \operatorname{Kern} \mathcal{S} \setminus \{0\}$. Of course, there exists three scalars $\alpha_i \neq 0$ such that $\alpha_i(\psi'_i)_{\partial\Omega} = 0$. Therefore, setting $\psi' = \alpha_i \psi'_i$, by Lemma 2 $v[\psi] = 0$ in $\mathbb{R}^2$ so that by (13) $\psi' = 0$ and, as a consequence, $\dim \mathfrak{C} \leq 2$.

If $\operatorname{Kern} \mathcal{S} = 2$, we have noting to prove. If $\operatorname{Kern} \mathcal{S} = \{0\}$, by Fredholm alternative, the problem

$$\mathcal{S}[\psi] = e_i, \tag{20}$$

($\{e_1, e_2\}$ orthonormal basis of $\mathbb{R}^2$) has a unique solution $\psi'_i \in L^2(\partial\Omega)$, $i = 1, 2$. If $\alpha_1 \psi'_1 + \alpha_2 \psi'_2 = 0$, then

$$\mathcal{S}[\alpha_1 \psi'_1 + \alpha_2 \psi'_2] = \alpha_1 e_1 + \alpha_2 e_2 = 0 \Rightarrow \alpha_1 = \alpha_2 = 0. \tag{21}$$

On the other hand, if $\psi' \in \mathfrak{C}$, then there exist scalars α_i such that $\mathcal{S}[\psi'] = \alpha_1 \mathcal{S}[\psi'_1] + \alpha_2 \mathcal{S}[\psi'_2]$. Then, setting $\psi = \psi' - \alpha_1 \psi'_1 + \alpha_2 \psi'_2$ and noting that $\int_{\partial\Omega} \psi = 0$, the field $v[\psi]$ is a solution to system (9) vanishing on $\partial\Omega$ and at infinity., Therefore by the uniqueness theorem and (13) we have that $\psi' = \alpha_1 \psi'_1 + \alpha_2 \psi'_2$. If $\operatorname{Kern} \mathcal{S} = \langle \alpha \psi'_1 \rangle_{\alpha \in \mathbb{R}}$, with $(\psi'_1)_{\partial\Omega} = e_1$ (say), we see that equation (20) has a unique solution ψ'_2. From (21) it follows $\alpha_2 = 0$ and from $\alpha_1 (\psi'_1)_{\partial\Omega} = 0$ that $\alpha_1 = 0$.

Let us note that the linear space

$$\left\{ \int_{\partial\Omega} \psi' : \psi' \in \mathfrak{C} \right\}$$

has dimension two.

3 Existence and uniqueness theorems

The results of the foregoing section allow us to make use of Fredholm's alternative to quickly prove existence for the displacement problems of linear elastostatics in the exterior domain (2)

$$\begin{aligned} \operatorname{div} \mathbb{C}[\nabla u] &= 0 \quad \text{in } \Omega, \\ u &= a \quad \text{on } \partial\Omega. \end{aligned} \tag{22}$$

where a is an assigned field on $\partial\Omega$.

Theorem 1. *Let Ω be a Lipschitz exterior domain of $\mathbb{R}^2$ and let $\mathbb{C}$ be constant, symmetric and positive definite. If $a \in L^2(\partial\Omega)$ and then there exists $\psi \in W^{-1,2}(\partial\Omega)$ such that the field*

$$u = v[\psi] + \gamma\kappa, \tag{23}$$

is a C^∞ solution to system (22), where $\gamma = \dim \operatorname{Kern} \mathcal{S}$, κ is defined by

$$\kappa \cdot \int_{\partial\Omega} \psi' = \int_{\partial\Omega} a \cdot \psi', \quad \forall \psi' \in \operatorname{Kern} \mathcal{S}, \tag{24}$$

$u \in W^{1/2,2}_{\text{loc}}(\overline{\Omega})$ and $u \xrightarrow{\text{nt}} a$. Moreover, if $a \in W^{1,2}(\partial\Omega)$, then $\psi \in L^2(\Omega)$ and $u \in W^{3/2,2}_{\text{loc}}(\overline{\Omega})$.

Proof. Consider the equation

$$\mathcal{S}[\psi] = a - \gamma\kappa$$

in $W^{-1,2}(\partial\Omega)$. Since by (27) $a - \gamma\kappa$ is orthogonal to every $\psi' \in \operatorname{Kern} \mathcal{S}$, existence immediately follows from Fredholm's alternative. The nontangential convergence of u to a as well as the regularity of u are a consequence of the representation (26).

In general, the solution in Theorem 1 behaves at infinity as the fundamental solution, *i.e*, $u(x) = O(\log r)$. In order to determine a solution to the problem

$$\begin{aligned} \operatorname{div} \mathbb{C}[\nabla u] &= 0 \quad \text{in } \Omega, \\ u &= a \quad \text{on } \partial\Omega, \\ \lim_{r\to+\infty} u(x) &= u_\infty, \end{aligned} \tag{25}$$

where u_∞ is an assigned constant vector, we use Lemma 4. Let u be the solution (26) and let $\psi' \in \mathfrak{C}$ be such that

$$(\psi + \psi')_{\partial\Omega} = 0.$$

Then, setting

$$\hat{\psi} = \psi + \psi', \quad u_\infty = \gamma\kappa - \mathcal{S}[\psi']$$

the field

$$\hat{u} = v[\hat{\psi}] + u_\infty, \tag{26}$$

is a solution to (25) provided

$$u_\infty \cdot \int_{\partial\Omega} \psi' = \int_{\partial\Omega} a \cdot \psi', \quad \forall\, \psi' \in \mathfrak{C}. \tag{27}$$

Therefore, the following theorem holds true.

Theorem 2. *Let Ω be a Lipschitz exterior domain of $\mathbb{R}^2$ and let $\mathbb{C}$ be constant, symmetric and positive definite. If $a \in L^2(\partial\Omega)$ satisfies* (27), *then problem* (25) *has a solution expressed by* (26).

Let us pass to show that the above solutions are unique in a *natural function class*. To this end, assume $a \in L^2(\partial\Omega)$ and denote by $\{a_k\}_{k\in\mathbb{N}}$ a sequence in $W^{1,2}(\partial\Omega)$ which converges strongly in $L^2(\partial\Omega)$ to a.

Let u_k be the solution to problem (22) with data (a_k, f), expressed by (23), and let z be the solution to the system (17), where $\varphi \in C_0^\infty(\Omega)$ satisfies

$$\int_\Omega \varphi = 0, \quad \int_\Omega \varphi \cdot v[\psi'] = 0, \quad \psi' \in \mathfrak{C}. \tag{28}$$

Taking into account that

$$u_k = O(\log r), \quad \nabla u_k = O(r^{-1}), \quad \nabla_j z = O(r^{-j}),\ j = 0, 1,$$

a simple computation yields

$$\int_\Omega \varphi \cdot u_k = \int_{\partial\Omega} a_k \cdot \mathbb{C}[\nabla z] n. \tag{29}$$

Then, letting $k \to +\infty$ in (29), we have

$$\int_\Omega \varphi \cdot u = \int_{\partial\Omega} a \cdot \mathbb{C}[\nabla z] n. \tag{30}$$

Of course, if $a = 0$, then $u \in \mathfrak{F}$, with

$$\mathfrak{F} = \{v[\psi'] - \mathcal{S}[\psi'],\ \psi' \in \mathfrak{C}\}, \quad \dim \mathfrak{F} = 2.$$

Following J. Nečas [9], a solution $u \in L^2_{\mathrm{loc}}(\overline{\Omega})$ to equation (30) is said to be a *very weak solution* to problem (22). Of course, u must satisfy (30) for every $\varphi \in C_0^\infty(\Omega)$ which meets (28), with z solution to system (17). Therefore we have

Theorem 3. *Let Ω be a Lipschitz exterior domain of $\mathbb{R}^2$ and let $\mathbb{C}$ be constant, symmetric and positive definite. If $a \in L^2(\partial\Omega)$, then problem* (22) *has a unique very weak solution expressed by* (23) *modulo a field in $\mathfrak{F}$.*

If u converges to the constant vector u_∞ at infinity, then, taking into account $(28)_1$ and that

$$\int_{\partial\Omega} \mathbb{C}[\nabla z]n = 0,$$

from (30) it easily follows (27). Then we can state

Theorem 4. *Let Ω be a Lipschitz exterior domain of $\mathbb{R}^2$ and let $\mathbb{C}$ be constant, symmetric and positive definite. If $a \in L^2(\partial\Omega)$, then problem* (25) *has a unique very weak solution expressed by* (26) *if and only if a and u_∞ satisfy the compatibility condition* (27).

Even if conditions (26) are necessary and sufficient for the existence of a solution u to system (25), they are not useful to select the boundary data assuring the desired existence, unless we are able to recover the fields of $\mathfrak{C}$. We aim at observing now that this is possible if the body is isotropic and $\partial\Omega$ is an ellipse.

Let the elastic body be also isotropic. Then the elasticity tensor takes the form [6]

$$\mathbb{C}[\nabla u] = 2\mu\left\{\hat{\nabla} u + \frac{\nu}{1-2\nu}(\operatorname{div} u)\mathcal{I}\right\} \tag{31}$$

and (9), (10) respectively become $(1)_1$ and

$$\mathcal{Q}(z) = \frac{1}{4(1-\nu)\mu}\left[(3-4\nu)\mathcal{U}(z)\mathcal{I} + \frac{z\otimes z}{\omega_n|z|^n}\right] \tag{32}$$

where μ is the shear modulus. Condition (8) is now equivalent to $\mu > 0$ and $\nu \in (-\frac{1}{2}, 1)$.

For $\nu = 1/2$, (32) gives the fundamental solution

$$\begin{aligned} \mathbb{S}(z) &= \left[\mathcal{U}(z)\mathcal{I} + \frac{z\otimes z}{\omega_n|z|^n}\right], \\ \mathcal{P}(z) &= -\nabla\mathcal{U}(z) \end{aligned} \tag{33}$$

to the Stokes system

$$\begin{aligned} \Delta u &= \nabla p \\ \operatorname{div} u &= 0. \end{aligned}$$

Let $\mathcal{S}_h[\psi]$ be the trace of the harmonic single layer potential $v_h[\psi]$ and let $\mathcal{S}_\sigma[\psi]$ be the trace of the Stokes single layer potential $v_\sigma[\psi]$ defined by (33). Let

$$\mathfrak{C}_{\sigma,0} = \{\psi : \mathcal{S}_\sigma[\psi] = \text{constant}\}, \quad \mathfrak{C}_{\sigma,0} = \{\psi : \mathcal{S}_h[\psi] = \text{constant}\}$$

The following result is classical. A simple proof can be found in [7] (Sect. 4) and [2].

Lemma 5. *Let $\mathbb{C}$ be expressed by* (31) *and let Ω be an ellipsoid of $\mathbb{R}^m$. Then*

$$\mathfrak{C}_{\sigma,0} = \operatorname{sp}\left\{[n(\xi)\cdot\xi]e_i\right\}_{i=1,2}, \quad \mathfrak{C}_{h,0} = \operatorname{sp}\left\{n(\xi)\cdot\xi\right\}$$

Since by (32)

$$v[\psi] = \lambda_1 v_h[\psi] + \lambda_2 v_\sigma[\psi],$$

with

$$\lambda_1 = \frac{1-2\nu}{2(1-\nu)\mu}, \quad \lambda_2 = \frac{1}{4(1-\nu)\mu},$$

from Lemma 5 it follows that

$$\mathfrak{C}_0 = \mathrm{sp}\left\{[n(\xi)\cdot\xi]e_i\right\}_{i=1,2}.$$

Therefore, in particular, we have the *Stokes paradox for the ellipse*

Theorem 5. *Let Ω be the exterior of an ellipse and let $\mu > 0$, $\nu \in (-\frac{1}{2}, 1)$. If $a \in L^1(\partial\Omega)$, then system* (1) *has a solution if and only if*

$$u_\infty = \frac{1}{2ab}\int_{\partial\Omega}(n\cdot\xi)a$$

Note that if $\partial\Omega$ is a disk, we recover the classical result of M. Picone (4).

References

[1] R.R. Coifman, A. McIntosh and I. Meyer: L'intégrale de Cauchy definit un operateur borné sur L^2 pour les courbes Lipschitziennes, *Ann. Math.***122**, 361–387 (1982).

[2] V. Coscia and R. Russo: Some remarks on the Dirichlet problem in plane exterior domains, *Ricerche Mat.* **53**, 31–41 (2007)

[3] E. B. Fabes: Layer potential methods for boundary value problems on Lipschitz domains, *Lectures Notes in Math.*, Springer–verlag **1344** (1988), 55–80.

[4] E.B. Fabes, M. Jodeit Jr. and N.M. Rivière: Potential techniques for boundary value problems on C^1 domains, *Acta Math.* **141** (1978), 165–185.

[5] G.P. Galdi and C. G. Simader, Existence, uniqueness and L^q estimates for the Stokes problem in an exterior domain, *Arch. Rational Mech. Anal.* **112** (1990), 291–318.

[6] M.E. Gurtin: The linear theory of elasticity, in *Handbuch der Physik* (ed. C. Truesedell) vol. VIa/2 (1972), Springer–Verlag.

[7] Maremonti, P., Russo, R. and Starita, G.: On the Stokes equations: the boundary value problem. *Quad. Mat.*

[8] C. Miranda: *Istituzioni di analisi funzionale lineare.* Unione Matematica Italiana, Oderisi Gubbio Editrice (1978).

[9] J. Nečas: *Les méthodes directes en théorie des équations élliptiques*, Masson-Paris and Academie-Prague (1967).

[10] M. Picone: Nuovi indirizzi di ricerca nella teoria e nel calcolo delle soluzioni di talune equazioni lineari alle derivate parziali della fisica–matematica, *Ann. Scuola Norm. Sup. Pisa* (2) **5** (1936), 213–288.

[11] R. Russo: On the existence of solution to the stationary Navier–Stokes equations, *Ricerche Mat.* (2003).

[12] G. Verchota: Layer potentials and regularity for the Dirichlet problem for Laplace's equation in Lipschitz domains, *J. Funct. Anal.* **59** (1984), 572–611.

Modelling of Capillary Forces in Viscous Fluids of Grade Two

A. Sellitto[*], *V. Triani*[†], *V.A. Cimmelli*[‡]

Department of Mathematics and Computer Science,
University of Basilicata,
Campus Macchia Romana, 85100, Potenza, Italy.

Dedicated to Professor Brian Straughan on the occasion of his 60th *birthday*

Abstract

We apply a generalized Liu procedure for the exploitation of second law to investigate the thermodynamic compatibility of the constitutive functions modelling capillarity forces in viscous fluids of grade two. We prove that these constitutive equations are fully compatible with the restrictions placed by the requirement of nonnegative entropy production.

Keywords: Capillary tubes, adhesion forces, cohesion forces, second grade fluids, extended Liu procedure.

1 Introduction

The evolution of fluids in capillary tubes is characterized by quasi-static state transitions. Moreover, the level of the fluid depends on its material properties. For instance, water tends to go up while mercury tends to go down. If the liquid surface has a supinates concavity, then the superficial strain is directed toward the center of the surface formed by the liquid, and so it has an upturn component. Due to this force, the liquid flows up until when the superficial strain and the gravitational force are in equilibrium. Such a phenomenon can be modelled by considering both the cohesion and the adhesion forces involving the molecules of the fluid. Due to the first type of forces, the molecules attract each other while, owing to the second type of forces, they interact with the molecules of another material as, for instance, the glass of a beaker in which the liquid is contained. We say that a fluid *wets* the surface of another material when the adhesion forces are bigger than the cohesion ones; in this case, the surface of a liquid column in a tube, such as water in glass, is concave and the fluid flows up into

[*] Email: `ant.sel@libero.it`
[†] Email: `vitrian@virgilio.it`
[‡] Email: `vito.cimmelli@unibas.it`

the capillary tube. If instead the adhesion forces are smaller than the cohesion ones, as for example mercury in glass, the surface of the liquid is convex and the level of the fluid in a system of capillary tubes diminishes proportionally to the decrement of the section of the tubes. Both adhesion and cohesion forces represent a typical example of long range interactions.

In order to model the effects due to long range interactions in fluids, in 1901 the Dutch physicist D. J. Korteweg formulated a constitutive equation for the Cauchy stress tensor including density gradients [1]. For the elastic part of the Cauchy stress tensor $\mathbf{T}$ he postulated a constitutive equation of the type

$$\mathbf{T} = \widehat{\mathbf{T}}\left(\varrho, \vartheta, \varrho_{,i}, \varrho_{,ij}\right), \tag{1}$$

where $\varrho(\mathbf{x},t)$ is the density of the fluid at the place $\mathbf{x}$ at time t, the symbol $(,k)$ means the partial derivative with respect to the coordinates x_k, $k = 1,2,3$, $\varrho_{,i}$ and $\varrho_{,ij}$ are, respectively, the components of the first and second spatial gradient of ϱ with respect to $\mathbf{x}$. The equation (1) describes the stress tensor in a special fluid of grade three. In fact, a material of grade N is such that the constitutive quantities depend not only on the first gradient of the deformation, namely the strain, but also on all gradients of the deformation less than or equal to the integer N, [2]-[5]. In the past three decades, higher grade materials have been employed not only to model capillarity effects but also to analyze the structure of liquid-vapor phase transitions [2]-[5], or to study pattern-forming instabilities in plastic deformations [6]. However, in order to make their constitutive equations compatible with second law of thermodynamics, the tenets of rational thermodynamics [7] have been changed in a subtle way. In fact, as it has been shown by Dunn and Serrin [8], if the classical balance equations for the wanted fields remain unchanged, then one obtains incompatibility with the requirement of nonnegative entropy production.

In order to solve this problem, these authors proposed a drastic modification of the local balance of energy by assuming the existence of an additional term, the so called interstitial work flux $\mathbf{u}$, which is engendered by long range interactions. Then, in the absence of heat sources, the local balance of energy reads:

$$\varrho\varepsilon_{,t} + \varrho\varepsilon_{,i}v_i - T_{ij}v_{i,j} + q_{i,i} - u_{i,i} = 0, \tag{2}$$

where ε represents the specific internal energy, $(,t)$ denotes the partial derivative with respect to time, q_i are the components of the heat flux vector. This generalized background makes third grade Korteweg fluids compatible with second law. Such a property has been proved by Dunn and Serrin by applying the classical Coleman-Noll procedure for the exploitation of second law [9]. However, the same methodology cannot be applied to study second grade Korteweg fluids, namely those fluids whose constitutive equations depend only on the first gradient of the mass density. In fact, by applying the Coleman-Noll procedure [9] one obtains that, if the constitutive functions depend on the gradients of the wanted fields up to the order N, then the thermodynamic potentials may only depend on the gradients up to the order $(N-1)$.

On the other hand, since the components of the stress tensor are given by suitable derivatives of the Helmholtz free energy $\psi = \varepsilon - \vartheta s$, where s denotes the specific entropy [8], for second grade fluids one gets $\mathbf{T}$ independent of the gradient of density. So, nonlocal effects cannot be modelled. Hence, a generalization of the classical procedures for the exploitation of second law of thermodynamics seems to be necessary.

In the present paper, we focus our attention on the celebrated Liu technique [10]. The classical Liu approach consists in forming an inequality, the Liu inequality, by adding to the local entropy production a linear combination of the balance equations for the unknown fields multiplied by suitable coefficients, namely the Lagrange multipliers [10]. A generalization of this procedure has been proposed by Cimmelli [11] on the example of rigid heat conductors. In such a case, an inequality is formed by adding to the entropy inequality a linear combination of the governing equations of the wanted fields and of their gradient extensions up to the order of the spatial derivatives entering the constitutive equations. The factors multiplying the field equations and their extensions are Lagrange multipliers. The thermodynamic restrictions on the constitutive functions are obtained by the observation that this inequality is linear in the highest derivatives[1] of the wanted fields, which are considered to be completely arbitrary. Here, we apply the procedure above to viscous, non heat conducting, second grade fluids. It is proved that this technique is capable to conserve the nonlocality of the constitutive equations.

2 Outline of the extended Liu procedure

Let us consider viscous, non heat conducting, second grade fluids. Their state space $\mathcal{Z}$ is given by

$$\mathcal{Z} = \{\varrho, \varrho_{,i}, \varepsilon, v_i, v_{i,j}\}. \tag{3}$$

For the sake of simplicity, we have substituted the absolute temperature ϑ with the specific internal energy ε. However, due to the positiveness of the specific heat $c = \dfrac{\partial \varepsilon}{\partial \vartheta}$ the two variables are in a one to one correspondence and hence there is no lack of generality.

The wanted fields entering $\mathcal{Z}$ are governed by the balances of mass, linear momentum and energy, i.e.

$$\varrho_{,t} + \varrho_{,i} v_i + \varrho v_{i,i} = 0, \tag{4}$$

$$\varrho v_{i,t} + \varrho v_{i,j} v_j + T_{ij,j} = 0, \tag{5}$$

$$\varrho \varepsilon_{,t} + \varrho \varepsilon_{,i} v_i - T_{ij} v_{i,j} + h_{i,i} = 0, \tag{6}$$

where h_i are the components of the net thermomechanical energy flux

$$\mathbf{h} = \mathbf{q} - \mathbf{u}. \tag{7}$$

Second law of thermodynamics forces the local entropy production to be nonnegative whatever the thermodynamic process is. This is equivalent to the unilateral differential constraint

$$\varrho s_{,t} + \varrho s_{,i} v_i + J_{i,i} = \varrho \sigma_s \geq 0, \tag{8}$$

where J_i are the components of the entropy flux vector and σ_s represents the specific entropy production. It is worth noticing that, although several representations of the entropy flux can be found in literature [7, 12, 13], here we do not assume any particular

[1] We call highest derivatives those time and space derivatives whose order is the highest one among them appearing into the entropy inequality.

constitutive equation for **J**, in order to make the class of the considered materials as large as possible. Taking into account (3), the inequality above can be written as

$$\begin{aligned}
&\varrho\left(\frac{\partial s}{\partial \varrho}\varrho_{,t}+\frac{\partial s}{\partial \varrho_{,j}}\varrho_{,jt}+\frac{\partial s}{\partial \varepsilon}\varepsilon_{,t}+\frac{\partial s}{\partial v_j}v_{j,t}+\frac{\partial s}{\partial v_{k,j}}v_{k,jt}\right)+\\
&+\varrho v_i\left(\frac{\partial s}{\partial \varrho}\varrho_{,i}+\frac{\partial s}{\partial \varrho_{,j}}\varrho_{,ji}+\frac{\partial s}{\partial \varepsilon}\varepsilon_{,i}+\frac{\partial s}{\partial v_j}v_{j,i}+\frac{\partial s}{\partial v_{k,j}}v_{k,ji}\right)+\\
&+\frac{\partial J_i}{\partial \varrho}\varrho_{,i}+\frac{\partial J_i}{\partial \varrho_{,j}}\varrho_{,ji}+\frac{\partial J_i}{\partial \varepsilon}\varepsilon_{,i}+\frac{\partial J_i}{\partial v_j}v_{j,i}+\frac{\partial J_i}{\partial v_{k,j}}v_{k,ji}\geq 0.
\end{aligned}\tag{9}$$

The extended equation for mass reads

$$\varrho_{,it}+\varrho_{,ji}v_j+\varrho_{,j}v_{j,i}+\varrho_{,i}v_{j,k}\delta_{jk}+\varrho v_{j,ki}\delta_{jk}=0.\tag{10}$$

Let us notice that the evolution equations (4) and (10) are independent of the form of the constitutive equations.

The evolution equation for the velocity gradient is

$$\begin{aligned}
&\varrho_{,i}v_{j,t}+\varrho v_{j,it}+\varrho_{,i}v_{j,k}v_k+\varrho v_{j,ki}v_k+\varrho v_{j,k}v_{k,i}-\\
&-\left(\frac{\partial^2 T_{jk}}{\partial \varrho^2}\varrho_{,i}+\frac{\partial^2 T_{jk}}{\partial \varrho_{,s}\partial \varrho}\varrho_{,si}+\frac{\partial^2 T_{jk}}{\partial \varepsilon\partial \varrho}\varepsilon_{,i}+\frac{\partial^2 T_{jk}}{\partial v_s\partial \varrho}v_{s,i}+\frac{\partial^2 T_{jk}}{\partial v_{s,z}\partial \varrho}v_{s,zi}\right)\varrho_{,k}-\\
&-\left(\frac{\partial^2 T_{jk}}{\partial \varrho\partial \varrho_{,s}}\varrho_{,i}+\frac{\partial^2 T_{jk}}{\partial \varrho_{,z}\partial \varrho_{,s}}\varrho_{,zi}+\frac{\partial^2 T_{jk}}{\partial \varepsilon\partial \varrho_{,s}}\varepsilon_{,i}+\frac{\partial^2 T_{jk}}{\partial v_z\partial \varrho_{,s}}v_{z,i}+\frac{\partial^2 T_{jk}}{\partial v_{z,m}\partial \varrho_{,s}}v_{z,mi}\right)\varrho_{,sk}-\\
&-\left(\frac{\partial^2 T_{jk}}{\partial \varrho\partial \varepsilon}\varrho_{,i}+\frac{\partial^2 T_{jk}}{\partial \varrho_{,s}\partial \varepsilon}\varrho_{,si}+\frac{\partial^2 T_{jk}}{\partial \varepsilon^2}\varepsilon_{,i}+\frac{\partial^2 T_{jk}}{\partial v_s\partial \varepsilon}v_{s,i}+\frac{\partial^2 T_{jk}}{\partial v_{s,z}\partial \varepsilon}v_{s,zi}\right)\varepsilon_{,k}-\\
&-\left(\frac{\partial^2 T_{jk}}{\partial \varrho\partial v_s}\varrho_{,i}+\frac{\partial^2 T_{jk}}{\partial \varrho_{,z}\partial v_s}\varrho_{,zi}+\frac{\partial^2 T_{jk}}{\partial \varepsilon\partial v_s}\varepsilon_{,i}+\frac{\partial^2 T_{jk}}{\partial v_z\partial v_s}v_{z,i}+\frac{\partial^2 T_{jk}}{\partial v_{z,m}\partial v_s}v_{z,mi}\right)v_{s,k}-\\
&-\left(\frac{\partial^2 T_{jk}}{\partial \varrho\partial v_{s,z}}\varrho_{,i}+\frac{\partial^2 T_{jk}}{\partial \varrho_{,m}\partial v_{s,z}}\varrho_{,mi}+\frac{\partial^2 T_{jk}}{\partial \varepsilon\partial v_{s,z}}\varepsilon_{,i}+\frac{\partial^2 T_{jk}}{\partial v_m\partial v_{s,z}}v_{m,i}+\frac{\partial^2 T_{jk}}{\partial v_{m,p}\partial v_{s,z}}v_{m,pi}\right)v_{s,zk}-\\
&-\frac{\partial T_{jk}}{\partial \varrho}\varrho_{,ki}-\frac{\partial T_{jk}}{\partial \varrho_{,s}}\varrho_{,ski}-\frac{\partial T_{jk}}{\partial \varepsilon}\varepsilon_{,ki}-\frac{\partial T_{jk}}{\partial v_s}v_{s,ki}-\frac{\partial T_{jk}}{\partial v_{s,z}}v_{s,zki}=0.
\end{aligned}\tag{11}$$

Now, a new general inequality can be formed by adding to (9) a linear combination of (4)-(6) and of the gradient extensions (10) and (11). The consequences of such an inequality will be derived in Section 3.

3 Restrictions placed by second law

Taking into account (4)-(6) and (10)-(11), the inequality (9) yields

$$\left(\varrho\frac{\partial s}{\partial \varrho}-\lambda^{\varrho}\right)\varrho_{,t}+\left(\varrho\frac{\partial s}{\partial \varrho_{,i}}-\lambda_i^{\varrho}\right)\varrho_{,it}+\varrho\left(\frac{\partial s}{\partial \varepsilon}-\lambda^{\varepsilon}\right)\varepsilon_{,t}+\left(\varrho\frac{\partial s}{\partial v_i}-\varrho\lambda_i^{\mathrm{v}}-\lambda_{il}^{\mathrm{v}}\varrho_{,l}\right)v_{i,t}+$$
$$+\varrho\left(\frac{\partial s}{\partial v_{i,l}}-\lambda_{il}^{\mathrm{v}}\right)v_{i,lt}+\lambda_{il}^{\mathrm{v}}\frac{\partial T_{ij}}{\partial \varrho_{,s}}\varrho_{,sjl}+\lambda_{il}^{\mathrm{v}}\frac{\partial T_{ij}}{\partial \varepsilon}\varepsilon_{,jl}+\lambda_{il}^{\mathrm{v}}\frac{\partial T_{ij}}{\partial v_{s,h}}v_{s,hjl}+$$
$$+f\left(\varrho,\varrho_{,i},\varrho_{,ij},\varepsilon,\varepsilon_{,i},v_i,v_{i,j},v_{i,jk},\lambda^{\varrho},\lambda_i^{\varrho},\lambda^{\varepsilon},\lambda_i^{\mathrm{v}},\lambda_{il}^{\mathrm{v}}\right)\geq 0, \tag{12}$$

where the quantities $\{\lambda^{\varrho},\lambda_i^{\varrho},\lambda^{\varepsilon},\lambda_i^{\mathrm{v}},\lambda_{il}^{\mathrm{v}}\}$ are Lagrange multipliers and f means a suitable regular scalar function. The terms entering the inequality (12) which are not included in f are linear with respect to the highest derivatives $\varrho_{,t}$, $\varrho_{,it}$, $\varrho_{,sjl}$, $\varepsilon_{,t}$, $\varepsilon_{,jl}$, $v_{i,t}$, $v_{i,lt}$ and $v_{s,hjl}$.

Since these derivatives are completely arbitrary and independent of their coefficients, should even one of these coefficients be different from zero, the inequality (12) could be easily violated.

As a consequence, they must vanish and this leads to the restrictions

$$\lambda^{\varrho}=\varrho\frac{\partial s}{\partial \varrho}, \tag{13}$$

$$\lambda_i^{\varrho}=\varrho\frac{\partial s}{\partial \varrho_{,i}}, \tag{14}$$

$$\lambda^{\varepsilon}=\frac{\partial s}{\partial \varepsilon}, \tag{15}$$

$$\lambda_i^{\mathrm{v}}=\frac{\partial s}{\partial v_i}-\frac{1}{\varrho}\lambda_{il}^{\mathrm{v}}\varrho_{,l}, \tag{16}$$

$$\lambda_{il}^{\mathrm{v}}=\frac{\partial s}{\partial v_{i,l}}, \tag{17}$$

$$\lambda_{kl}^{\mathrm{v}}\frac{\partial T_{ki}}{\partial \varrho_{,j}}=0, \tag{18}$$

$$\lambda_{kl}^{\mathrm{v}}\frac{\partial T_{ki}}{\partial \varepsilon}=0, \tag{19}$$

$$\lambda_{jl}^{\mathrm{v}}\frac{\partial T_{ji}}{\partial v_{k,t}}=0, \tag{20}$$

$$f\left(\varrho,\varrho_{,i},\varrho_{,ij},\varepsilon,\varepsilon_{,i},v_i,v_{i,j},v_{i,jk},\lambda^{\varrho},\lambda_i^{\varrho},\lambda^{\varepsilon},\lambda_i^{\mathrm{v}},\lambda_{il}^{\mathrm{v}}\right)\geq 0. \tag{21}$$

Furthermore, by a rather cumbersome analysis [2], it is possible to prove that the inequality (21) leads to the following additional relations

$$\varrho v_l\frac{\partial s}{\partial \varrho_{,k}}+\frac{\partial J_l}{\partial \varrho_{,k}}-\lambda^{\varepsilon}\frac{\partial h_l}{\partial \varrho_{,k}}+\lambda_i^{\mathrm{v}}\frac{\partial T_{il}}{\partial \varrho_{,k}}+\lambda_{il}^{\mathrm{v}}\frac{\partial^2 T_{ij}}{\partial \varrho\partial \varrho_{,k}}\varrho_{,j}-\lambda_l^{\varrho}v_k+$$
$$-\lambda_{ij}^{\mathrm{v}}\frac{\partial^2 T_{il}}{\partial \varrho_{,k}\partial \varrho}\varrho_{,j}+\lambda_{il}^{\mathrm{v}}\frac{\partial T_{ik}}{\partial \varrho}+\lambda_{ij}^{\mathrm{v}}\frac{\partial^2 T_{il}}{\partial \varrho_{,k}\partial v_t}v_{t,j}+\lambda_{il}^{\mathrm{v}}\frac{\partial^2 T_{ij}}{\partial v_s\partial \varrho_{,k}}v_{s,j}=0, \tag{22}$$

[2]Such an analysis has been developed in detail in [14] for a more general class of fluids.

$$
\begin{aligned}
&\frac{\partial J_j}{\partial \varepsilon} - \lambda^{\varepsilon}\frac{\partial h_j}{\partial \varepsilon} + \lambda_i^{\mathrm{v}}\frac{\partial T_{ij}}{\partial \varepsilon} + \lambda_{ij}^{\mathrm{v}}\frac{\partial^2 T_{il}}{\partial \varrho \partial \varepsilon}\varrho_{,l} + \\
&+\lambda_{il}^{\mathrm{v}}\frac{\partial^2 T_{ij}}{\partial \varepsilon \partial \varrho}\varrho_{,l} + \lambda_{il}^{\mathrm{v}}\frac{\partial^2 T_{ij}}{\partial \varepsilon \partial v_t}v_{t,l} + \lambda_{ij}^{\mathrm{v}}\frac{\partial^2 T_{il}}{\partial v_s \partial \varepsilon}v_{s,l} = 0,
\end{aligned}
\tag{23}
$$

$$
\begin{aligned}
&\varrho v_l\frac{\partial s}{\partial v_{h,k}} + \frac{\partial J_l}{\partial v_{h,k}} - \varrho\lambda_l^{\varrho}\delta_{hk} - \lambda^{\varepsilon}\frac{\partial h_l}{\partial v_{h,k}} + \lambda_i^{\mathrm{v}}\frac{\partial T_{il}}{\partial v_{h,k}} - \varrho\lambda_{hl}^{\mathrm{v}}v_k + \\
&+\lambda_{il}^{\mathrm{v}}\frac{\partial^2 T_{ij}}{\partial \varrho \partial v_{h,k}}\varrho_{,j} + \lambda_{il}^{\mathrm{v}}\frac{\partial^2 T_{ij}}{\partial v_s \partial v_{h,k}}v_{s,j} + \lambda_{il}^{\mathrm{v}}\frac{\partial T_{ik}}{\partial v_h} + \lambda_{ij}^{\mathrm{v}}\frac{\partial^2 T_{il}}{\partial v_{h,k}\partial \varrho}\varrho_{,j} + \\
&+\lambda_{ij}^{\mathrm{v}}\frac{\partial^2 T_{il}}{\partial v_{h,k}\partial v_s}v_{s,j} = 0.
\end{aligned}
\tag{24}
$$

It is worth observing that by equations (18)-(20) and (22)-(24) follows a rather general thermodynamic framework, such that the Cauchy stress tensor depends on all the gradients of the wanted fields too. So the extended technique applied here dismantles new physical properties and preserves the nonlocality of the constitutive equations also for second grade fluids.

Let us notice that, once the Lagrange multipliers given by equations (13)-(17) are substituted into the restrictions (18)-(20) and (22)-(24), one obtains the same thermodynamic restrictions presented in [15], which have been proved through an extension of the classical Coleman-Noll technique.

This result represents a further progress on the problem of the equivalence of classical and extended Liu and Coleman-Noll procedures for the exploitation of second law. A proof of the equivalence of the classical methods for rigid heat conductors has been obtained in [16] while, for the class of fluids considered here, the proof of the equivalence has been carried out in [15].

4 Conclusions

A generalized Liu procedure for the exploitation of second law of thermodynamics has been applied to investigate the thermodynamic compatibility of the constitutive functions modelling second grade fluids. The main result is that these constitutive equations are fully compatible with the restrictions placed by the requirement of nonnegative entropy production. Such a compatibility, cannot be proved by applying the classical Coleman-Noll or Liu procedures. There are several physical systems and critical phenomena, such as liquid crystals [17], damage in solids [18, 19] or diffusive-hyperbolic heat conduction [20, 21], in which nonlocal potentials seems to be necessary. In all these situations, the technique we presented here can give a more deep insight into the physical properties of the materials at hand.

Acknowledgements

Vita Triani thanks for financial support the *Regione Basilicata* POR 2000-2006, Research Project *Modelli matematici in termodinamica di non equilibrio.*

References

[1] D. J. Korteweg. *Sur la forme qui prennent les equations du mouvement des fluids si l'on tient compte des forces capillaires par des variations de densité.* Arch. Néer. Sci. Exactes, Ser II., **6** (1901), 1-24.

[2] E. C. Aifantis, J. Serrin. *The mechanical theory of fluid interfaces and Maxwell's rule.* J. Colloid Interface Sci., **96** (1983), 517-529.

[3] E. C. Aifantis, J. Serrin. *Equilibrium solutions in the mechanical theory of fluid microstructures.* J. Colloid Interface Sci., **96** (1983), 530-547.

[4] M. Slemrod. *Admissibility criteria for propagating phase boundaries in a van der Waals fluid.* Arch. Rat. Mech. Anal., **81** (1983), 301-315.

[5] M. Slemrod. *Dynamic phase transitions in a van der Waals fluid.* J. Differential Equations, **52** (1984), 1-23.

[6] E. C. Aifantis. *Pattern formation in plasticity.* Int. J. Engng Sci., **33** (1995), 2161-2178.

[7] B. D. Coleman, D. R. Owen. *A mathematical foundation for thermodynamics.* Arch. Rat. Mech. Anal., **54** (1974), 1-104.

[8] J. E. Dunn, J. Serrin. *On the thermomechanics of the interstitial working.* Arch. Rat. Mech. Anal., **88** (1985), 95-133.

[9] B. D. Coleman, W. Noll. *The thermodynamics of elastic materials with heat conduction and viscosity.* Arch. Rat. Mech. Anal., **13** (1963), 167-178.

[10] I-Shi Liu. *Method of Lagrange multipliers for exploitation of the entropy principle.* Arch. Rat. Mech. Anal., **46** (1972), 131-148.

[11] V. A. Cimmelli. *An extension of Liu procedure in weakly nonlocal thermodynamics.* J. Math. Phys., **48** (2007), 113510 (13 pages).

[12] J. Verhás. *On the entropy current.* J. Non-Equilib. Thermodyn., **8** (1983), 201-206.

[13] B. Nyíri. *On the entropy current.* J. Non-Equilib. Thermodyn., **16** (1991), 179-186.

[14] V. A. Cimmelli, A. Sellitto, V. Triani. *A new thermodynamic framework for second grade fluids.* Forthcoming.

[15] V. Triani, A. Sellitto, V. A. Cimmelli. *Exploitation of second law for second grade fluids: generalized Coleman-Noll and Liu procedures in comparison.* Proceedings NTFSM-08, 2008, in this volume.

[16] V. Triani, C. Papenfuss, V. A. Cimmelli, W. Muschik. *Exploitation of the Second Law: Coleman-Noll and Liu Procedure in Comparison.* J. Non-Equilib. Thermodyn., **33** (2008), 47-60.

[17] C. Papenfuss, S. Forest. *Thermodynamical framework for higher grade material theories with internal variables or additional degrees of freedom.* J. Non-Equilib. Thermodyn., **31** (2006), 319-353.

[18] Z. P. Bazant, G. Pijauder-Cabot. *Nonlocal damage theory.* J. Eng. Mech. ASCE, **113** (1987), 1512-1533.

[19] P. Ván, B. Vásárhelyi. *Second Law of thermodynamics and the failure of rock materials.* In: Rock Mechanics in the National Interest, Vol.1, J. P. Tinucci, D. Elsworth and K. A. Heasley Eds, Balkema Publishers, Lisse, 2001, 767-773.

[20] V. A. Cimmelli, W. Kosiński. *Nonequilibrium semi-empirical temperature in materials with thermal relaxation.* Arch. Mech., **43** (1991), 753-767.

[21] V. A. Cimmelli, P. Ván, *The effects of nonlocality on the evolution of higher order fluxes in nonequilibrium thermodynamics.* J. Math. Phys. **46** (2005), 112901 (15 pages).

Absorbing Sets of the Positive Solutions of the Schnackenberg Reaction-Diffusion System under Neumann Homogeneous Boundary Conditions

S. Rionero, *A. Tataranni*†

Università di Napoli *Federico II*,
Dipartimento di Matematica e Applicazioni *R. Caccioppoli*,
Complesso Universitario Monte S. Angelo, via Cinzia, 80126 Napoli, Italia.

Abstract

The positive solutions of a binary reaction-diffusion system of P.D.Es modelling the Schnackenberg chemical autocatalytic reaction are considered. The absorbing sets are obtained.

1 Introduction

The reaction diffusion system of P.D.Es are often used for modelling the chemical reactions [1]-[7]. Here we consider the initial boundary value problem

$$\begin{cases} U_t = \gamma(a - U + U^2V) + \Delta U \\ \\ V_t = \gamma(b - U^2V) + d\Delta V \end{cases} \tag{1}$$

$$U(x,0) = U_0 \geq 0 \quad V(x,0) = V_0 \geq 0, \quad \forall x \in \Omega \tag{2}$$

$$\nabla U \cdot \mathbf{n} = 0, \nabla V \cdot \mathbf{n} = 0, \qquad \forall (x,t) \in \partial\Omega \times \mathbb{R}^+ \tag{3}$$

in a fixed open bounded domain $\Omega \subset \mathbb{R}^3$, with C^{2+p}, $p > 0$, boundary and a, b, γ, d positive constants such that :

$$a + b > 0, \, b > 0, \, d > 0. \tag{4}$$

System (1)-(3) contains as particular case the system introduced by Schnackenberg for trimolecular autocatalytic reactions [1]. It is easily seen that (1)-(3) admits the steady state

$$U_* = a + b, \quad V_* = \frac{b}{(a+b)^2}. \tag{5}$$

*Email: rionero@unina.it
†Email: assunta.tataranni@dma.unina.it

In two previous papers [8]-[9] the stability-instability of (5) under Dirichlet boundary conditions has been studied.
Here we consider the positive solutions of (1)-(3) with the aim of showing the definitively boundness of solutions. Precisely, after some preliminaries (Section 2), in Section 3 we determine the absorbing sets.

2 Preliminaries

Let $< \cdot, \cdot >$ and $\| \cdot \|$ denote the scalar product and the $L_2(\Omega)$-norm respectively.
For the existence of positive solutions of (1)-(3), belonging to $L_2(\Omega)$, we refer to [10].
We here recall that the spectral problem

$$\begin{cases} \Delta\phi + \lambda\phi = 0, \quad \nabla\phi \neq 0 \quad \text{in}\,\Omega \\ \nabla\phi \cdot \mathbf{n} = 0, \quad \text{on}\,\partial\Omega \end{cases} \tag{6}$$

with $\phi \in W^{1,2}(\Omega)$, admits a lowest positive eigenvalue $\overline{\alpha}$ given by

$$\overline{\alpha} = \min_{W^{1,2}(\Omega)} \frac{\|\nabla\phi\|^2}{\|\phi\|^2} \tag{7}$$

(where $\mathbf{n}$ is the outward unit normal to $\partial\Omega$) and that

$$\int_\Omega \phi \, d\Omega = 0 \tag{8}$$

for any eigenfunction of (6). Further the spectral problem

$$\begin{cases} \Delta\phi + \lambda\phi = 0, \quad \text{in}\,\Omega \\ \nabla\phi \cdot \mathbf{n} = 0, \quad \text{on}\,\partial\Omega \end{cases} \tag{9}$$

admits a complete orthogonal sequence of eigenfunctions $\{\Phi_n\}$,$(n = 0, 1, 2, \ldots)$, with

$$\Phi_0 = \text{const} \neq 0 \tag{10}$$

and $\Phi_i, i \leq 1$, eigenfunction of (6). The associate sequence of eigenvalues of (9) verifies the conditions

$$\alpha_0 = 0,\ 0 < \overline{\alpha} = \alpha_1 < \alpha_2 < \alpha_3 \ldots \tag{11}$$

Then any smooth function $\Phi \in W^{1,2}(\Omega)$ verifying (6_2) can be developable in the Fourier series

$$\Phi = c + \sum_1^n a_i \Phi_i \tag{12}$$

with $c = \text{const}$. We assume that (12) is derivable term by term as needed and setting

$$U = U_* + u, \quad V = V_* + v \tag{13}$$

the perturbation (u, v) to the steady state verifies

$$\int_\Omega u \, d\Omega = \int_\Omega v \, d\Omega = 0 \tag{14}$$

3 Absorbing sets

Let $|\Omega|$ =measure of Ω.
Then -in view of (8) and (12)- it turns out that

$$< V, \Delta V >=< V_*, \Delta V > + < v, \Delta v >\leq -\overline{\alpha}\|v\|^2. \tag{15}$$

Since

$$\|V\| = \|(V - V_*) + V_*\| \leq \|V - V_*\| + \|V\| \tag{16}$$

implies

$$\|V - V_*\|^2 \geq (\|V\| - \|V_*\|)^2 \geq \|V\|^2 - 2\|V_*\|\|V\| \tag{17}$$

it one obtains

$$\|v\|^2 \geq \|V\|^2 - 2\|V_*\|\|V\| \tag{18}$$

and hence

$$< V, \Delta V >\leq -\overline{\alpha}\|V\|^2 + 2\overline{\alpha}\|V_*\|\|V\|. \tag{19}$$

In view of (1)-(3) one obtains

$$\frac{1}{2}\frac{d}{dt}\|V\|^2 \leq \gamma b < 1, V > +(d-1) < V, \Delta V > + < V, \Delta V > \tag{20}$$

and hence

$$\frac{1}{2}\frac{d}{dt}\|V\|^2 \leq \gamma b|\Omega|^{1/2}\|V\| - \overline{\alpha}(d-1)\|V\|^2 + 2\overline{\alpha}(d-1)\|V_*\|\|V\| - \|\nabla V\|^2. \tag{21}$$

Setting

$$W = U + V \tag{22}$$

one obtains

$$\begin{cases} \dfrac{\partial W}{\partial t} = \gamma(a+b) - \gamma W + \gamma V + \Delta W + (d-1)\Delta V \quad \text{in } \Omega \\ \\ \nabla W \cdot \mathbf{n} = 0 \qquad \text{on } \partial\Omega \end{cases} \tag{23}$$

and hence

$$\begin{aligned} \frac{1}{2}\frac{d}{dt}\|W\|^2 &= \gamma(a+b) < 1, W > -\gamma\|W\|^2 + \gamma < V, W > + < W, \Delta W > + \\ &+(d-1) < W, \Delta V > \end{aligned} \tag{24}$$

By virtue of

$$\begin{cases} < 1, W >= |\Omega|^{1/2}\|W\| \\ \\ < V, W >\leq \|V\|\|W\| \\ \\ < W, \Delta W >= -\|\nabla W\|^2 \\ \\ < W, \Delta V >= - < \nabla V, \nabla W >\leq \|\nabla V\|\|\nabla W\| \end{cases} \tag{25}$$

it turns out that

$$\frac{1}{2}\frac{d}{dt}\|W\|^2 \leq \gamma(a+b)|\Omega|^{1/2}\|W\| - \gamma\|W\|^2 + \gamma < V, W > -\|\nabla W\|^2 + \\ + \frac{d-1}{2\varepsilon}\|\nabla V\|^2 + \frac{\varepsilon(d-1)}{2}\|\nabla W\|^2 \tag{26}$$

with ε a positive constant to be chosen later.
By virtue of (20) and (26) one obtains

$$\frac{1}{2}\frac{d}{dt}(\|W\|^2 + \|V\|^2) \leq [\gamma b + 2\overline{\alpha}(d-1)V_*]|\Omega|^{1/2}\|V\| - \overline{\alpha}(d-1)\|V\|^2 + \gamma(a+b)|\Omega|^{1/2}\|W\| + \\ -\gamma\|W\|^2 + \frac{\gamma}{2\eta}\|V\|^2 + \frac{\gamma\eta}{2}\|W\|^2 - \left(1 - \varepsilon\frac{d-1}{2}\right)\|\nabla W\|^2 - \left(1 - \frac{d-1}{2\varepsilon}\right)\|\nabla V\|^2 \tag{27}$$

Now, on choosing ε in a such a way that

$$\frac{d-1}{2} < \varepsilon < \frac{2}{d-1},$$

i.e. for $1 < d < 3$, it follows that

$$\frac{1}{2}\frac{d}{dt}(\|W\|^2 + \|V\|^2) \leq a_1|\Omega|^{1/2}(\|V\| + \|W\|) - \left[\overline{\alpha}(d-1) - \frac{\gamma}{2\eta}\right]\|V\|^2 \\ -\gamma\left(1 - \frac{\eta}{2}\right)\|W\|^2 \tag{28}$$

where

$$a_1 = \max\{\gamma b + 2\overline{\alpha}(d-1)V_*,\ \gamma(a+b)\}$$

, and η is a suitable positive constant.
Choosing

$$\frac{\gamma}{2\overline{\alpha}(d-1)} < \eta < 2,$$

for $\gamma < 4\overline{\alpha}(d-1)$ one obtains

$$\frac{1}{2}\frac{d}{dt}(\|W\|^2 + \|V\|^2) \leq a_1|\Omega|^{1/2}(\|V\| + \|W\|) - \delta(\|V\|^2 + \|W\|^2) \tag{29}$$

with

$$\delta = \inf\left\{\overline{\alpha}(d-1) - \frac{\gamma}{2\eta}, \gamma(1-\eta/2)\right\}. \tag{30}$$

Finally, from (29) it easily follows that

$$\frac{d}{dt}\left(\|W\|^2 + \|V\|^2\right)^{1/2} \leq \sqrt{2}a_1|\Omega|^{1/2} - \delta\left(\|W\|^2 + \|V\|^2\right)^{1/2} \tag{31}$$

and so, setting

$$R = \sqrt{2}a_1|\Omega|^{1/2}$$

the following theorem holds

Theorem 1. *Let*

$$\begin{cases} 1 < d < 3 \\ \gamma < 4\overline{\alpha}(d-1). \end{cases} \tag{32}$$

Then, measuring in the phase-plane the distance of (U, V) *from the origin by*

$$E = \sqrt{\|W\|^2 + \|V\|^2} \tag{33}$$

, it turns out that for any $\overline{\varepsilon} > 0$*, the ball* $S(\overline{\varepsilon})$ *centered at the origin* $(0,0)$ *defined by*

$$S(\overline{\varepsilon}) = \left\{(U,V) : E < (1+\overline{\varepsilon})\frac{R}{\delta}\right\} \tag{34}$$

is an absorbing set.

Proof. In fact, let

$$E_0 = (\|W_0\|^2 + \|V_0\|^2)^{1/2} < (1+\overline{\varepsilon})\frac{R}{\delta}. \tag{35}$$

Then

$$\left(\frac{dE}{dt}\right)_{t=0} = \frac{d}{dt}(\|W_0\|^2 + \|V_0\|^2)^{1/2} \leq R - \frac{\delta(1+\overline{\varepsilon})R}{\delta} = -\overline{\varepsilon}R < 0 \tag{36}$$

i.e. $S(\varepsilon)$ is invariant.
Denoting by I a bounded set of the phase-space then exists a positive constant $M(I)$ such that

$$(U,V) \in I \Rightarrow E < M. \tag{37}$$

Therefore, since (31) implies

$$E \leq E_0 e^{-\delta t} + \frac{R}{\delta} \tag{38}$$

for any $t > \overline{t}$ with

$$Me^{-\delta \overline{t}} + \frac{R}{\delta} = (1+\varepsilon)\frac{R}{\delta} \tag{39}$$

i.e.

$$\overline{t} = \frac{1}{\delta}\log\frac{M\delta}{\varepsilon} \tag{40}$$

any trajectory, initially in I, belongs to $S(\varepsilon)$.

Acknowledgements

This paper has been performed under the auspices of the G.N.F.M. – INDAM and the PRIN project: Waves and Stability in Continuous Media – 2007. A. Tataranni thanks to contract n° 21 – MAT/07 – *Questioni di stabilità-instabilità non lineare per sistemi binari di P.D.Es in presenza di reazione-diffusione causata da problemi chimici*, (2008) Università *Federico II*, Naples, Dipartimento di Matematica ed Applicazioni *R. Caccioppoli.*

References

[1] Schnackenberg, J. (1979) *Simple chemical reaction systems with limit cycle behaviour*, J. Theor. Biol., **81**, 389-400.

[2] Turing, A.M. (1952) *The chemical basis of morphogenesis*, Phil. Trans. R. Soc. B, **237**, 37-72

[3] Dillon, R. Maini, P.K. and Othmer, H.G. (1994)*Pattern formation in generalized Turing systems*, J. Math. Biol., **32**, 345-393.

[4] Madzvamuse, A. Thomas, R.D.K. Maini, P.K. and Wathen, A.J.(2002) *A numerical approach to the study of spatial pattern formation in the ligaments of arcoid bivalves*, Bull. Math. Biol., **64**, 501-530.

[5] Murray, J.D. (2003) *Mathematical Biology.I. An introduction*, Third edition,Interdisciplinary Applied Mathematics **17**, Springer Verlag, New York.

[6] Murray, J.D. (2003) *Mathematical Biology.II. Spatial model and Biomedical Applications*, Third edition,Interdisciplinary Applied Mathematics **18**, Springer Verlag, New York.

[7] Segel, L. and Jackson, J. (1972) *Dissipative structure: an explanation and an ecological example*, J. Theor. Biol., **37**, 545-559.

[8] Gentile M., Rionero S., Tataranni A. (2006) *On the stability of a binary chemical reaction diffusion system with Turing instability*. Rend. Acc. Sc. fi. mat. Napoli, **LXXIII**, 481-490,

[9] Gentile M., Tataranni A. *On nonlinear stability for a chemical reaction-diffusion system* Proceedings of XIV International Conference on Waves and Stability in Continuous Media (to appear)

[10] Mora X. (1983) *Semilinear parabolic problems define semiflows on C^k spaces* Trans. Amer. Math. Soc, **278**:1, 21-55

On the Deformation of the Surface of an Elastic Medium with a Cavity

*V. Tibullo**, *V. Zampoli*†

Department of Information Engineering and Applied Mathematics,
University of Salerno
Via Ponte don Melillo, 84084 Fisciano (SA), Italy

Abstract

The problem concerning the investigation about the anti-plane deformation of an half-plane surface, in presence of a cavity and a shear force on the external surface, is formulated and solved in some details. The analytical results obtained are used to solve numerically the problem for different configurations of the system. Finally, some graphics showing such solutions are reported.

1 Introduction

More studies have been developed to determine the boundary displacement in the context of the direct static anti-plane problem, in presence of defects or cavities. Recently, the inverse problem, always considered in the above mentioned frame, has been studied also [1]. The mathematical formulation of this (direct) problem is configured through the Boundary Integral Equation Method (BIEM) or, in a recent terminology, Boundary Element Method (BEM), applied to Laplace equation (see, for example, [2]).

Moreover, the detailed structure of Green's function for two-dimensional problems, both in the anti-plane (as in the present work) and in-plane cases, can be found in [3].

Advanced applications of Boundary Element Method, in the context of the theory of elastic solids, are given in [4]. A recent development of this theory is presented in [5].

It should be noted that the classical theory of the Boundary Element Method, or Boundary Integral Equation Method, briefly reviewed here, is currently still under a rapid development. An extensive review of the recent results about the above mentioned theory can be found in [6 − 15].

In this paper we study the deformation of the limit surface of an isotropic and homogeneous elastic medium, containing a cavity (in various positions) with circular and elliptic boundary, and subject to a shear force applied on the above mentioned limit surface.

*Email: vtibullo@unisa.it
†Email: vzampoli@unisa.it

We will develop this study in a half-plane context and considering the anti-plane case.

In section two, we will define the mathematical formulation of the problem;

in section three, we will study the solution of the problem;

in section four, we will construct the Green's function for the Laplace equation in the half-plane;

in section five, using the results of section four, we will investigate on the problem through an integral equation;

in section six, we will show some numerical examples and finally,

in section seven, we will discuss about the most important results obtained for cavities with different shapes, dimensions and positions, also showing several graphics.

2 Problem formulation

Let us consider a homogeneous and isotropic elastic medium filling the half space $y \geq 0$, and suppose the medium has a cavity, whose boundary is a cylinder with generators parallel to the z axis and whose section S with the plane $z = 0$ is delimited by a simple and closed piecewise regular curve $L = \partial S$.

Let the elastic medium be subject to a shear force per surface unit on the boundary, given by

$$\mathbf{t}(x, z) = \sigma_0 \delta (x - x_0) \mathbf{k} \tag{1}$$

where σ_0 is a positive constant, δ is the Dirac delta generalized function, and $\mathbf{k}$ is the z axis unit vector.

We want to study the direct static anti-plane problem, consisting in the determination of the displacement field $\mathbf{u}$ on the surface $y = 0$, assuming that shape, dimension and position of the cavity are known, and assuming that $\mathbf{u}$ has only the z component not vanishing (anti-plane problem):

$$\mathbf{u}(x, y, z) = w(x, y) \mathbf{k}. \tag{2}$$

Note that, for the system symmetry, (invariance respect to translation along the z axis), the function w cannot depend on the z coordinate, so from now on we will consider the relative 2-dimensional problem.

From the equation of motion for $\mathbf{u}$:

$$\rho \ddot{\mathbf{u}} = \mu \Delta \mathbf{u} + (\lambda + \mu) \nabla (\nabla \cdot \mathbf{u}) + \rho \mathbf{b} \tag{3}$$

(ρ is the mass density, Δ is the Laplace operator, $\mathbf{b}$ the body force field and λ, μ the Lamé elastic parameters) in absence of body forces and in the static case, we find that the equation becomes

$$\mu \Delta \mathbf{u} + (\lambda + \mu) \nabla (\nabla \cdot \mathbf{u}) = \mathbf{0}$$

but from (2) we see that

$$\nabla \cdot \mathbf{u} = 0$$

so we obtain, for the unknown function w:

$$\Delta w(x, y) \equiv \frac{\partial^2 w}{\partial x^2} + \frac{\partial^2 w}{\partial y^2} = 0 \tag{4}$$

in the closed half-plane $Y = \{(x, y) \in R^2 : y \geq 0\}$.

If we introduce the usual *strain* and *stress* tensors $\mathbf{E}$ and $\mathbf{T}$, where

$$\mathbf{E} = \frac{1}{2}(\nabla\mathbf{u} + \nabla\mathbf{u}^T)$$

and

$$\mathbf{T} = \lambda \operatorname{tr}(\mathbf{E})\,\mathbf{I} + 2\mu\mathbf{E}, \tag{5}$$

we have

$$\mathbf{t} = \mathbf{Tn} = \mu\frac{\partial \mathbf{u}}{\partial \mathbf{n}} = \mu\frac{\partial w}{\partial \mathbf{n}}\mathbf{k} \tag{6}$$

where $\mathbf{n}$ represents the outward normal unit vector in the considered point of the boundary.

The third component of the stress vector is

$$t_3 = \mu\frac{\partial w}{\partial \mathbf{n}}$$

so the boundary conditions (of Neumann type) on w are

$$\mu\frac{\partial w}{\partial \mathbf{n}} = \mathbf{0} \tag{7}$$

for $(x, y) \in L$, where there is no surface force, and

$$\mu\frac{\partial w}{\partial \mathbf{n}} = \mu\frac{\partial w}{\partial y} = \sigma_0\delta\,(x - x_0) \tag{8}$$

on the external boundary ∂Y.

3 Solution of the problem

To solve the direct problem, we decompose the function w into the sum of two components

$$w\,(x, y) = \varphi^0\,(x, y) + \varphi\,(x, y)\,, \tag{9}$$

where φ^0 is assumed to be the solution of the problem in absence of the cavity, while φ represents the effect of the cavity on w.

Let us examine the equations these two functions must satisfy and the boundary condition we have to impose on them. Obviously, φ^0 must satisfy the same equation and the same boundary condition on ∂Y as w:

$$\begin{cases} \Delta\varphi^0\,(x, y) = 0 & \text{fory} > 0 \\ \mu\dfrac{\partial\varphi^0}{\partial y} = \sigma_0\delta\,(x - x_0) & \text{fory} = 0 \end{cases} \tag{10}$$

and there is no condition on L (remember that φ^0 is the solution in absence of the cavity). For φ we obtain

$$\Delta\varphi = \Delta\left(w - \varphi^0\right) = \Delta w - \Delta\varphi^0 = 0 \tag{11}$$

and

$$\frac{\partial \varphi}{\partial y} = \frac{\partial (w - \varphi^0)}{\partial y} = \frac{\partial w}{\partial y} - \frac{\partial \varphi^0}{\partial y} = 0 \qquad \text{for} y = 0$$

$$\frac{\partial \varphi}{\partial \mathbf{n}} = \frac{\partial (w - \varphi^0)}{\partial \mathbf{n}} = \frac{\partial w}{\partial \mathbf{n}} - \frac{\partial \varphi^0}{\partial \mathbf{n}} = -\frac{\partial \varphi^0}{\partial \mathbf{n}} \qquad \text{for (x, y)} \in \text{L}. \tag{12}$$

Finally, we have

$$\begin{cases} \Delta \varphi (x, y) = 0 & \text{for} y > 0 \\ \mu \dfrac{\partial \varphi}{\partial y} = 0 & \text{for} y = 0 \\ \mu \dfrac{\partial \varphi}{\partial \mathbf{n}} = -\mu \dfrac{\partial \varphi^0}{\partial \mathbf{n}} & \text{for (x, y)} \in \text{L}. \end{cases} \tag{13}$$

A solution for φ^0, as can be seen by direct substitution, is the following:

$$\varphi^0(x, y) = -\frac{\sigma_0}{2\pi\mu} \int_{-\infty}^{+\infty} \frac{e^{-i\alpha(x-x_0)-|\alpha|y}}{|\alpha|} d\alpha. \tag{14}$$

This solution can be obtained using a method analogous to the one used in the following section to find the Green's function for the Laplace equation in the half plane. Moreover, we have to note that the integral appearing in this solution does not converge, so, in order to have a more concrete solution, we can take the first partial derivatives of this function with respect to x, y, and then turn back to the function from its derivative. In this way we can write the solution as

$$\varphi^0(x, y) = \frac{\sigma_0}{2\pi\mu} \ln \left((x - x_0)^2 + y^2 \right). \tag{15}$$

We note that this solution is still divergent, but only in the point of application of the shear force, and this divergence can be interpreted as a consequence of the infinite force applied (along all the z axis). Again, this expression for φ^0 is not the same as (14); they differ for a constant value, as permitted by (13), but this constant should be infinite.

4 Green's function for the Laplace equation in the half-plane

In order to proceed in the search for a solution for φ, we need the Green's function $\Phi(\xi, \eta; x, y)$ for the Laplace equation in the half-plane, with a homogeneous boundary condition:

$$\begin{cases} \Delta \Phi (\xi, \eta; x, y) = -\delta (\xi - x) \delta (\eta - y) & \text{for} \eta > 0 \\ \dfrac{\partial \Phi}{\partial \eta} = 0 & \text{for} \eta = 0 \end{cases} \tag{16}$$

where (x, y) is a fixed point in the half-plane.

As first step, we consider the Fourier transform of the equation respect to the variable ξ, so that derivation respect to this variable becomes, apart from a factor $-i$, a multiplication by the new variable s in the Fourier transform domain:

$$\frac{\partial}{\partial \xi} \Rightarrow -is \tag{17}$$

so, taking into account that the Fourier transform of a δ-Dirac function is an exponential function, as can be easily seen using its definition, the equation becomes a linear ordinary differential equation:

$$\frac{d^2\tilde{\Phi}}{d\eta^2} - s^2\tilde{\Phi}(s,\eta) = -e^{isx}\delta(\eta - y) \tag{18}$$

where s is considered as a parameter and $\tilde{\Phi}$ is the Fourier transform of the function Φ with respect to the variable ξ, defined as

$$\tilde{\Phi}(s,\eta) = \int_{-\infty}^{+\infty} \Phi(\xi,\eta)\, e^{is\xi} d\xi. \tag{19}$$

The solution of the homogeneous equation associated with (18) is the following:

$$\tilde{\Phi}_0(s,\eta) = A(s)\, e^{-|s|\eta} + B(s)\, e^{+|s|\eta}. \tag{20}$$

Now, let us find a particular solution to (18) through the Fourier transform of this equation with respect to η. We will use the symbol $\hat{\Phi}$ to denote the Fourier transform of $\tilde{\Phi}$ with respect to η, and t will be the variable conjugate to η:

$$-t^2\hat{\Phi}(s,t) - s^2\hat{\Phi}(s,t) = -e^{isx}e^{ity} \tag{21}$$

from which directly follows

$$\hat{\Phi}(s,t) = \frac{e^{isx}e^{ity}}{s^2+t^2}. \tag{22}$$

From the last equation and taking into account the relation (19), we obtain

$$\tilde{\Phi}_{\text{part}}(s,\eta) = \frac{1}{2\pi}\int_{-\infty}^{+\infty} \frac{e^{isx}e^{ity}}{s^2+t^2} e^{-it\eta} dt = \frac{e^{isx}e^{-|s||\eta-y|}}{2|s|} \tag{23}$$

where the integral can be easily calculated using standard complex analysis methods.

We are now able to write the full solution of (18):

$$\tilde{\Phi}(s,\eta) = A(s)\, e^{-|s|\eta} + B(s)\, e^{+|s|\eta} + \frac{e^{isx}e^{-|s||\eta-y|}}{2|s|}. \tag{24}$$

To find out the unknown functions A, B we can use the boundary (homogeneous) conditions on $\tilde{\Phi}$. First of all, this solution must be bounded when $\eta \to \infty$, and this implies

$$B(s) = 0. \tag{25}$$

In the second place, using the boundary condition $(16)_2$ near $\eta = 0$ we have

$$\frac{d\tilde{\Phi}}{d\eta}(s,0) = -|s|A(s) + \frac{1}{2}e^{isx}e^{-|s|y} = 0 \tag{26}$$

from which we obtain

$$A(s) = \frac{1}{2|s|}e^{isx}e^{-|s|y}. \tag{27}$$

Finally, the solution of (16) is

$$\Phi(\xi,\eta;x,y) = \frac{1}{4\pi}\int\limits_{-\infty}^{+\infty} \frac{e^{-|s|(\eta+y)} + e^{-|s||\eta-y|}}{|s|} e^{-is(\xi-x)}ds. \tag{28}$$

Also for this expression, we can obtain a simpler form, differing from the preceding by an (infinite) constant:

$$\Phi(\xi,\eta;x,y) = -\frac{1}{4\pi}\left[\ln\left((\xi-x)^2+(\eta-y)^2\right)+\ln\left((\xi-x)^2+(\eta+y)^2\right)\right]. \tag{29}$$

It should be noted that an alternative derivation of the just constructed Green's function can be carried out on the basis of the so-called *mirror* (or *virtual*) *image* method (see [3, 5]). However, the method applied in the present work is more general, and can be used also for different geometries, not only for the half-plane.

5 Reduction to an integral equation

We can show now, using the result of the last section, that φ must satisfy an integral equation.

We start from writing the following integral

$$\int\limits_{\Omega} (\varphi\Delta\Phi - \Phi\Delta\varphi)\, d\xi d\eta \tag{30}$$

where Ω is the closure of the set $Y - S$. Using the divergence theorem, we obtain, from the last integral, the following relation

$$\int\limits_{\Omega} (\varphi\Delta\Phi - \Phi\Delta\varphi)\, d\xi d\eta = \int\limits_{\partial\Omega} \left(\varphi\frac{\partial\Phi}{\partial\mathbf{n}} - \Phi\frac{\partial\varphi}{\partial\mathbf{n}}\right) dl \tag{31}$$

whereas, with the aid of (13) and (16), we obtain

$$\int\limits_{\Omega} (\varphi\Delta\Phi - \Phi\Delta\varphi)\, d\xi d\eta = -\varphi(x,y). \tag{32}$$

So, having

$$\partial\Omega = \partial Y - L \tag{33}$$

(L, as part of $\partial\Omega$, is travelled clockwise), we obtain

$$\varphi(x,y) = \int_L \left(\varphi \frac{\partial \Phi}{\partial \mathbf{n}} - \Phi \frac{\partial \varphi}{\partial \mathbf{n}}\right) dl = \int_L \left(\varphi \frac{\partial \Phi}{\partial \mathbf{n}} + \Phi \frac{\partial \varphi^0}{\partial \mathbf{n}}\right) dl. \tag{34}$$

Using the results of the potential theory, we have

$$\lim_{(x,y)\to(X,Y)} \int_L \varphi \frac{\partial \Phi}{\partial \mathbf{n}} (\xi, \eta; x, y)\, dl = \frac{\varphi(X,Y)}{2} + \int_L \varphi \frac{\partial \Phi}{\partial \mathbf{n}} (\xi, \eta; X, Y)\, dl \tag{35}$$

where $(X, Y) \in L$, so we can write the (34), using the above limit, as

$$\frac{1}{2}\varphi(X,Y) - \int_L \varphi(\xi,\eta) \frac{\partial \Phi}{\partial \mathbf{n}} (\xi, \eta, X, Y)\, dl = f(X,Y) \tag{36}$$

where

$$f(X,Y) = \int_L \frac{\partial \varphi^0}{\partial \mathbf{n}} (\xi, \eta)\, \Phi(\xi, \eta, X, Y)\, dl. \tag{37}$$

The integral equation (36) contains a curvilinear integral. Expressing the curve L through a parametric representation

$$\begin{cases} \xi = \xi(t) \\ \eta = \eta(t) \end{cases} \qquad a \leq t \leq b \tag{38}$$

and introducing the notations

$$\begin{cases} \varphi_L(t) = \varphi(\xi(t), \eta(t)) \\ K(s,t) = \dfrac{\partial \Phi}{\partial \mathbf{n}} (\xi(t), \eta(t), \xi(s), \eta(s)) \sqrt{(\xi'(t))^2 + (\eta'(t))^2} \\ f_L(s) = f(\xi(s), \eta(s)) \end{cases} \tag{39}$$

where $s \in [a, b]$ is the value of the parameter for which $(\xi(s), \eta(s)) = (X, Y)$, we can rewrite the (36) in the following form

$$\frac{1}{2}\varphi_L(s) - \int_a^b K(s,t)\varphi_L(t)dt = f_L(s). \tag{40}$$

We can see that the (40) is a second kind non homogeneous Fredholm integral equation that, once fixed the curve L and calculated the known term $f_L(s)$, can be solved using standard numerical algorithms.

6 Numerical examples

In this section we want to show some applications of the results obtained. We have to solve numerically the integral equation specific of the problem for different shapes, dimensions and positions of the cavity.

The solution of the integral equation is obtained using the co-location method. This method consists basically in considering the integral equation, valid in an interval, only in a discrete and finite sequence of equally spaced points of such interval:

$$\frac{1}{2}\varphi_L(t_i) - \int_a^b K(t_i, t)\,\varphi_L(t)\,dt = f_L(t_i) \qquad i = 1, \ldots, n \tag{41}$$

with

$$s_i = a + \varepsilon(i - 1/2) \qquad i = 1, \ldots, n \tag{42}$$

and with

$$\varepsilon = \frac{b - a}{n} \tag{43}$$

where n is a positive integer conveniently chosen. The integral is then written as a sum of integrals on each of the small intervals $[s_i - \varepsilon/2, s_i + \varepsilon/2]$ in which the whole interval is divided, and supposing, using an approximation, the unknown function constant in each interval:

$$\frac{1}{2}\varphi_L(t_i) - \sum_{j=1}^{n} \varphi_L(t_j) \int_{t_j-\varepsilon/2}^{t_j+\varepsilon/2} K(t_i, t_j)\,dt = f_L(t_i) \qquad i = 1, \ldots, n. \tag{44}$$

As can be easily understood, this last set of equations is a linear system with respect to variables $\varphi_L(t_i)$ and, once solved this system and substituted t to t_i, equation (44) can be conveniently used as an interpolation formula to obtain $\varphi_L(t)$ for each desired $t \in [a, b]$.

7 Conclusions

In the following figures, some solutions for the function φ, depending on the cavity features, are shown.

In particular, we consider two different kinds of cavity, one of circular shape, with radius $r = 1$, and one of elliptic shape, with semiaxis $a = 3$, $b = 1$. In this context, we identify the first as the *circular* case and the second as the *elliptic* case. In each one of the four following diagrams, a comparison between the same kind of cavity, but with different positions, is made.

Circular case.

In figure 1, a comparison between φ for a circular cavity, whose centre coordinate ξ_0 is always zero, while its depth is respectively $\eta_0 = 3, 6, 9$, is shown.

As can be easily deduced from diagrams, a more superficial cavity has a greater influence on φ, but it is important to notice that the dependence of φ on the depth of the cavity is non-linear, as the figure 5 of $\varphi_{\max}$ versus η_0 shows. On the other side, in figure 2, the dependence of φ on an horizontal change of the cavity centre position ($\xi_0 = 0, 3, 6$; $\eta_0 = 3$) is shown: proceeding in the cavity shift with the same step as the case shown in figure 1, it is easy to notice that the influence of the cavity position on φ value results much more linear (see figure 6).

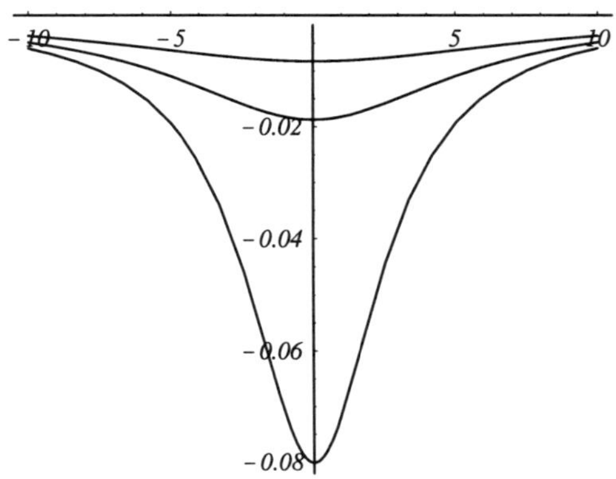

Figure 1: $\varphi(\xi, 0)$ for various depths of a circular cavity.

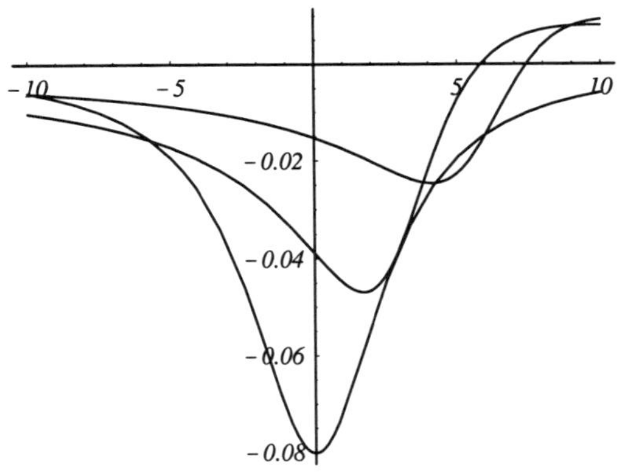

Figure 2: $\varphi(\xi, 0)$ for various horizontal shifts of a circular cavity.

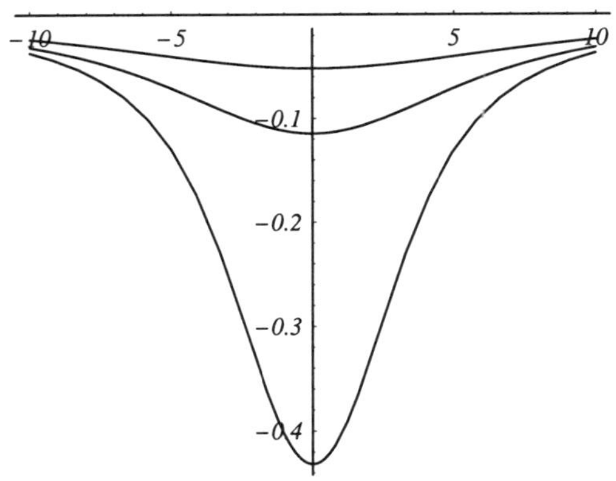

Figure 3: $\varphi(\xi, 0)$ for various depths of an elliptic cavity.

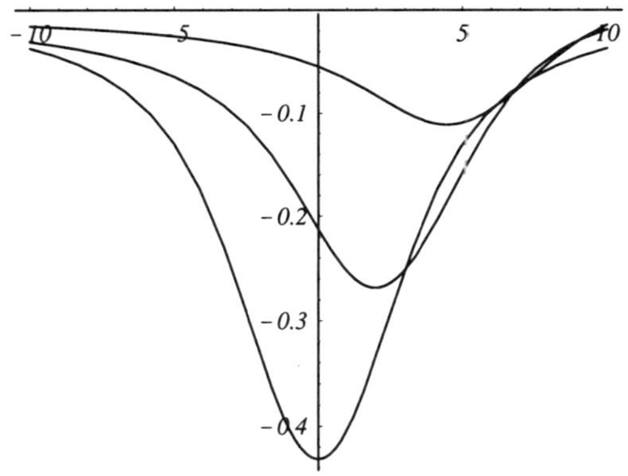

Figure 4: $\varphi(\xi, 0)$ for various horizontal shifts of an elliptic cavity.

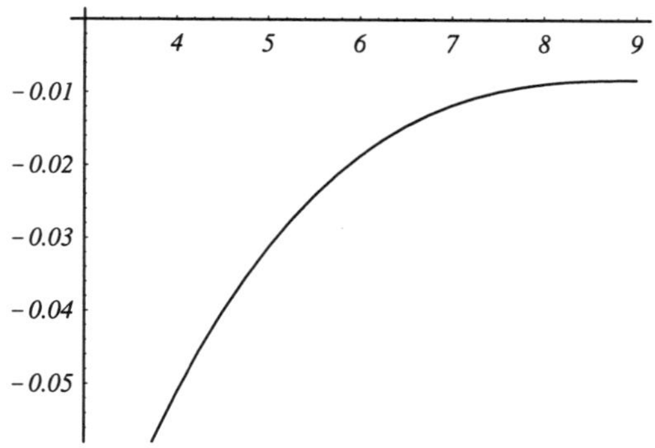

Figure 5: $\varphi_{\max}$ for various depths of a circular cavity.

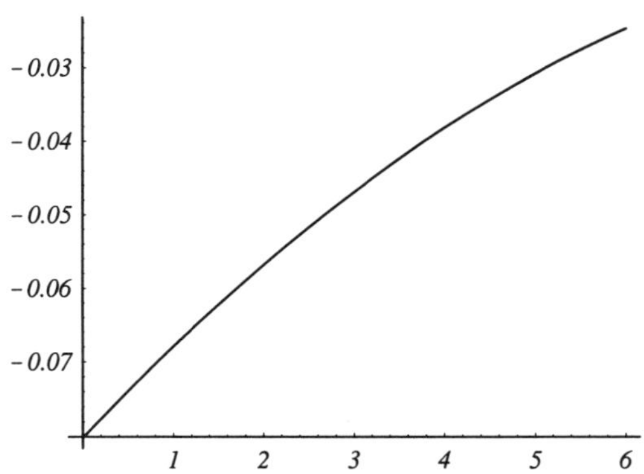

Figure 6: $\varphi_{\max}$ for various horizontal shifts of a circular cavity.

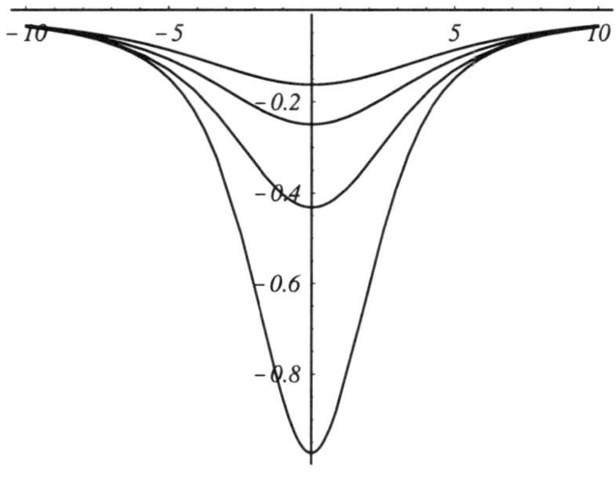

Figure 7: $\varphi(\xi, 0)$ for various vertical shifts of an elliptic cavity.

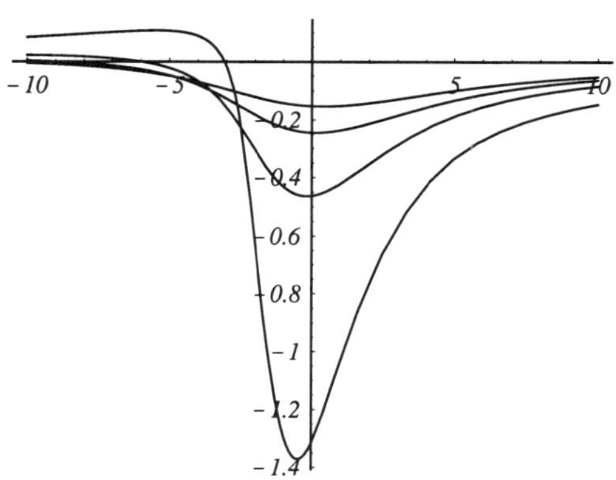

Figure 8: $\varphi(\xi, 0)$ for various vertical shifts of a rotated elliptic cavity.

Elliptic case.

Figures 3 and 4 show, respectively, the effect on φ due to a vertical and horizontal change of cavity position, exactly in the same way as the circular case, but with the elliptic cavity above mentioned. It is absolutely interesting the fact that, while the relationship between vertical and horizontal case is nearly the same as the circular case, in the elliptic case φ values are in general greater. In particular, φ value in the elliptic case results, for the same cavity position, about five times greater than in the circular case.

Finally, in figures (7) and (8) we point out:

- the effect on φ of the above mentioned elliptic cavity, but considered much nearer to the surface (η_0 starts from 2);
- the effect on φ of the same elliptic cavity, considered at the same depths, but inclined of an angle $\theta = \pi/6$ respect to the horizontal axis.

While the fact that a higher and not inclined cavity has a greater influence on φ could be expected, it is interesting to note that, rotating it, φ becomes naturally not vertically symmetric anymore, and the influence of the rotation on the module of φ becomes greater as the cavity comes near to the surface.

References

[1] M. Ciarletta, G. Iovane, M. A. Sumbatyan, Inverse Problem on Crack Reconstruction in the Elastic Half-Space: Anti-Plane Case, preprint DIIMA n. 03/2005, submitted to Inverse Problems in Science and Engineering, 2004.

[2] P. K. Banerjee, R. Butterfield, Boundary Element Methods in Engineering Science, McGraw-Hill, London, 1981.

[3] C. A. Brebbia, J. C. F. Telles, L. C. Wrobel, Boundary Element Techniques, Springer-Verlag, Berlin, 1984.

[4] S. L. Crouch, A. M. Starfield, Boundary Element Methods in Solid Mechanics, Allen & Unwin, London, 1983.

[5] M. Bonnet, Boundary Integral Equations Methods for Solids and Fluids, John Wiley, New York, 1999.

[6] G. O. Ribeiro, J. P. Cruz, T. S. A. Ribeiro, A. B. Jorge, T. A. Cruse, Evaluation of non-singular BEM algorithms for potential problems, Engineering Analysis with Boundary Elements, accepted for publication, 2003.

[7] A. B. Jorge, G. O. Ribeiro, T. A. Cruse, T. S. Fisher, Self-regular boundary integral equation formulations for Laplace's equation in 2-D, International Journal for Numerical Methods in Engineering, Vol. 51, No. 1, 1-30, 2001.

[8] L. Gaul, M. Kogl, M. Wagner, Boundary Element Methods for Engineers and Scientists, 488 pages, ISBN 3-540-00463-7, Springer-Verlag, Berlin, 2003.

[9] M. H. Aliabadi, The Boundary Element Method, Vol. 2, Applications in Solids and Structures, 598 pages, ISBN 0-470-84298-9, John Wiley & Sons, Inc., 2002.

[10] Ch. Zhang, A. Savaidis, 3-D Transient Dynamic Crack Analysis by a Novel Time-Domain BEM, Computer Modeling in Engineering & Sciences, Vol. 4, No. 5, 603-618, 2003.

[11] K. Park, P. K. Banerjee, Two and three-dimensional transient thermoelastic analysis by BEM via particular integrals, International Journal of Solids and Structures, Vol. 39, 2871-2892, 2002.

[12] K. Park, P. K. Banerjee, Two and three-dimensional soil consolidation by BEM via particular integrals, Computer Methods in Applied Mechanics and Engineering, Vol. 191, 3233-3255, 2002.

[13] Y. J. Liu, H. Fan, Boundary Integral Equation Formulations for Piezoelectric Solids Containing Thin Shapes, Electronic Journal of Boundary Elements, Vol. BETEQ 2001, No. 1, 12-19, 2002.

[14] D. Z. Y. Liu, F. Rizzo, Nearly singular and hypersingular integrals in the boundary element method: in C. Brebbia et al., Boundary Elements XV.Worcester, MA: Computational Mechanics Publications, 1993.

[15] M. Denda, The Green's function BEM for bimaterial solids applied to edge stress concentration problems, Rutgers University, Mech. & Aero. Engng. Dept. Piscataway, NJ 08854-8058, U.S.A. M.Denda / Electronic Journal of Boundary Elements, Vol. 1, No. 2, 112-131, 2003.

Exploitation of Second Law for Second Grade Fluids: Generalized Coleman-Noll and Liu Procedures in Comparison

V. Triani[*] *A. Sellitto*[†] *V.A. Cimmelli*[‡]

Department of Mathematics and Computer Science,
University of Basilicata,
Campus Macchia Romana, 85100, Potenza, Italy.

Dedicated to Professor Brian Straughan on the occasion of his 60th *birthday*

Abstract

A thermodynamic model of nonlocal viscous fluids is developed. The restrictions placed by second law are derived by applying a generalization of the classical Coleman-Noll procedure. It is proved that the same results can be obtained by applying a generalization of the classical Liu procedure.

Keywords: Weakly nonlocal thermodynamics, second grade fluids, extended Coleman-Noll procedure, extended Liu procedure.

1 Introduction

In order to model capillarity effects in fluids, in 1901 D. J. Korteweg introduced density gradients in the elastic part of the Cauchy stress tensor $\mathbf{T}$[1]. He postulated the constitutive equation

$$\mathbf{T} = \widehat{\mathbf{T}}\left(\varrho, \vartheta, \varrho_{,i}, \varrho_{,ij}\right), \tag{1}$$

where $\varrho(\mathbf{x}, t)$ is the density of the fluid at the place $\mathbf{x}$ at time t, the symbol $(, k)$ means the partial derivative with respect to the coordinates x_k, $k = 1, 2, 3$, $\varrho_{,i}$ and $\varrho_{,ij}$ are, respectively, the components of the first and second spatial gradient of ϱ with respect to $\mathbf{x}$. Eq. (1) describes the stress tensor in a special fluid of grade three. In fact, according to the modern terminology, a material of grade N is such that the constitutive quantities are permitted to depend not only on the first gradient of the deformation, the strain, but also on all gradients of the deformation less than or equal

*Email: vitrian@virgilio.it
†Email: ant.sel@libero.it
‡Email: vito.cimmelli@unibas.it

to the integer N. In recent years, such higher grade materials have been employed not only to model capillarity effects but also to analyze the structure of liquid-vapour phase transitions under both static [2, 3] and dynamic [4, 5] conditions. It was observed by Dunn and Serrin [6] that these higher grade models are, in general, incompatible with the restrictions placed on (1) by second law of thermodynamics. In order to find a remedy to such an incompatibility, the authors proposed a generalization of the classical local balance of energy. Indeed, they postulated the existence of a rate of supply of mechanical energy, the so called interstitial working, such that the local balance of energy takes the form

$$\varrho\varepsilon_{,t} + \varrho\varepsilon_{,i}v_i - T_{ij}v_{i,j} + q_{i,i} - u_{i,i} = 0, \tag{2}$$

where ε represents the specific internal energy, the symbol $(,t)$ denotes the partial derivative with respect to time, q_i are the components of the heat flux vector and u_i are the components of the interstitial work flux. The introduction of the interstitial work flux $\mathbf{u}$ changes the expression of the local energy flux that in conventional thermodynamics is given by the heat flux $\mathbf{q}$ while now is given by

$$\mathbf{h} = \mathbf{q} - \mathbf{u}. \tag{3}$$

By substituting (2) into the Clausius-Duhem inequality [7]

$$\varrho s_{,t} + \varrho s_{,i}v_i + \left(\frac{q_i}{\vartheta}\right)_{,i} \geq 0, \tag{4}$$

wherein s is the specific entropy, one obtains the following entropy inequality

$$\varrho\left(\varepsilon_{,t} + \varepsilon_{,i}v_i - \vartheta s_{,t} - \vartheta s_{,i}v_i\right) - T_{ij}v_{i,j} - u_{i,i} + \frac{1}{\vartheta}q_i\vartheta_{,i} \leq 0. \tag{5}$$

Then, by applying the classical Coleman-Noll procedure for exploitation of second law [8], Dunn and Serrin were able to prove that the constitutive equation (1) is compatible with the thermodynamic restrictions imposed by (5).

However, the method proposed by Dunn and Serrin cannot be applied to second grade fluids, namely to those fluids for which the constitutive equations take the form

$$F = \tilde{F}\left(\varrho, \vartheta, \varrho_{,i}\right). \tag{6}$$

In fact, by the classical Coleman-Noll procedure it follows that if the constitutive equations depend on the gradients of the unknown fields up to the order N, then the thermodynamic potentials s and ε may only depend on the gradients up to the order $(N-1)$. Since for the class of fluids represented by (6) the components of the stress tensor are given by suitable derivatives of the Helmholtz free energy $\psi = (\varepsilon - \vartheta s)$ [6], one obtains that the stress tensor cannot depend on the density gradient, i.e. nonlocal effects cannot be modelled. The considerations above prove that, in dealing with second grade materials, the classical Coleman-Noll procedure is not sufficient to conserve all the necessary nonlocalities into the constitutive equations. Hence, a generalization of such a procedure seems to be necessary.

In the present paper such a generalization is achieved. To this end, we substitute into the dissipation inequality the balance laws of the wanted fields together with their gradient extensions. The thermodynamic restrictions follow by an analytical procedure starting by the observation that, in order to make satisfied the entropy inequality for all solutions of the system of balance equations, the coefficients of the highest derivatives[1] of the state functions have to be zero. It is proved that this technique is capable to conserve the nonlocality of the constitutive equations also for second grade fluids.

It is worth noticing that, for the same class of fluids the compatibility with second law can be also investigated by applying a generalized Liu procedure [9]. Such an investigation has been achieved in [10] and in [11] for a wider class of fluids. In both cases, we applied the same methodology already developed in [12] for a rigid heat conductor.

As a further result, in the present paper we prove that both techniques are equivalent, since they lead to the same set of thermodynamic restrictions. The equivalence of the classical procedures was already proved in [13] on the example of rigid heat conductors. To our opinion, the present result represents a further confirmation that it is possible to prove a general theorem of equivalence between the extended Coleman-Noll and Liu procedures by considering a deformable body whose constitutive equations contain space and time derivatives of the basic variables up to an arbitrary order.

2 An extended Coleman-Noll procedure

Let us consider a viscous, non heat conducting, second grade fluid. Its state space $\mathcal{Z}$ is spanned by the mass density ϱ, the specific internal energy ε, the material velocity $\mathbf{v}$ and the first order gradients of the mass density and of the velocity, namely

$$\mathcal{Z} = \{\varrho, \varrho_{,i}, \varepsilon, v_i, v_{i,j}\}. \tag{7}$$

For the sake of simplicity, we have substituted the absolute temperature ϑ with the specific internal energy ε. However, due to the positiveness of the specific heat $c = \dfrac{\partial \varepsilon}{\partial \vartheta}$ the two variables are in a one to one correspondence and hence there is no lack of generality.

The basic state variables are governed by the balances of mass, linear momentum and energy, i.e.

$$\varrho_{,t} + \varrho_{,i} v_i + \varrho v_{i,i} = 0, \tag{8}$$

$$\varrho v_{i,t} + \varrho v_{i,j} v_j - T_{ij,j} = 0, \tag{9}$$

$$\varrho \varepsilon_{,t} + \varrho \varepsilon_{,i} v_i - T_{ij} v_{i,j} + h_{i,i} = 0. \tag{10}$$

Now, in order to consider the extension of the Coleman-Noll method we have also to calculate the evolution equations for the gradients of the basic variables.
The extended equation for the mass reads

$$\varrho_{,it} + \varrho_{,ji} v_j + \varrho_{,j} v_{j,i} + \varrho_{,i} v_{j,k} \delta_{jk} + \varrho v_{j,ki} \delta_{jk} = 0. \tag{11}$$

Let us notice that the form of the evolution equations (8) and (11) is independent of the form of the constitutive equations.

[1]We call highest derivatives those time and space derivatives whose order is the highest one among them appearing into the entropy inequality.

The evolution equation for the gradient of velocity, is a little bit more complicated. It reads

$$
\begin{aligned}
&\varrho_{,i}v_{j,t}+\varrho v_{j,it}+\varrho_{,i}v_{j,k}v_k+\varrho v_{j,ki}v_k+\varrho v_{j,k}v_{k,i}-\\
&-\left(\frac{\partial^2 T_{jk}}{\partial\varrho^2}\varrho_{,i}+\frac{\partial^2 T_{jk}}{\partial\varrho_{,s}\partial\varrho}\varrho_{,si}+\frac{\partial^2 T_{jk}}{\partial\varepsilon\partial\varrho}\varepsilon_{,i}+\frac{\partial^2 T_{jk}}{\partial v_s\partial\varrho}v_{s,i}+\frac{\partial^2 T_{jk}}{\partial v_{s,z}\partial\varrho}v_{s,zi}\right)\varrho_{,k}-\\
&-\left(\frac{\partial^2 T_{jk}}{\partial\varrho\partial\varrho_{,s}}\varrho_{,i}+\frac{\partial^2 T_{jk}}{\partial\varrho_{,z}\partial\varrho_{,s}}\varrho_{,zi}+\frac{\partial^2 T_{jk}}{\partial\varepsilon\partial\varrho_{,s}}\varepsilon_{,i}+\frac{\partial^2 T_{jk}}{\partial v_z\partial\varrho_{,s}}v_{z,i}+\frac{\partial^2 T_{jk}}{\partial v_{z,m}\partial\varrho_{,s}}v_{z,mi}\right)\varrho_{,sk}-\\
&-\left(\frac{\partial^2 T_{jk}}{\partial\varrho\partial\varepsilon}\varrho_{,i}+\frac{\partial^2 T_{jk}}{\partial\varrho_{,s}\partial\varepsilon}\varrho_{,si}+\frac{\partial^2 T_{jk}}{\partial\varepsilon^2}\varepsilon_{,i}+\frac{\partial^2 T_{jk}}{\partial v_s\partial\varepsilon}v_{s,i}+\frac{\partial^2 T_{jk}}{\partial v_{s,z}\partial\varepsilon}v_{s,zi}\right)\varepsilon_{,k}-\\
&-\left(\frac{\partial^2 T_{jk}}{\partial\varrho\partial v_s}\varrho_{,i}+\frac{\partial^2 T_{jk}}{\partial\varrho_{,z}\partial v_s}\varrho_{,zi}+\frac{\partial^2 T_{jk}}{\partial\varepsilon\partial v_s}\varepsilon_{,i}+\frac{\partial^2 T_{jk}}{\partial v_z\partial v_s}v_{z,i}+\frac{\partial^2 T_{jk}}{\partial v_{z,m}\partial v_s}v_{z,mi}\right)v_{s,k}-\\
&-\left(\frac{\partial^2 T_{jk}}{\partial\varrho\partial v_{s,z}}\varrho_{,i}+\frac{\partial^2 T_{jk}}{\partial\varrho_{,m}\partial v_{s,z}}\varrho_{,mi}+\frac{\partial^2 T_{jk}}{\partial\varepsilon\partial v_{s,z}}\varepsilon_{,i}+\frac{\partial^2 T_{jk}}{\partial v_m\partial v_{s,z}}v_{m,i}+\frac{\partial^2 T_{jk}}{\partial v_{m,p}\partial v_{s,z}}v_{m,pi}\right)v_{s,zk}-\\
&-\frac{\partial T_{jk}}{\partial\varrho}\varrho_{,ki}-\frac{\partial T_{jk}}{\partial\varrho_{,s}}\varrho_{,ski}-\frac{\partial T_{jk}}{\partial\varepsilon}\varepsilon_{,ki}-\frac{\partial T_{jk}}{\partial v_s}v_{s,ki}-\frac{\partial T_{jk}}{\partial v_{s,z}}v_{s,zki}=0.
\end{aligned}
\tag{12}
$$

It is worth observing that both the tensorial order of the derivatives of the state functions and the number of the terms entering the extended balances increases very fast. This is characteristic of the extended methods.

The constitutive equations are restricted by second law of thermodynamics which reads

$$
\begin{aligned}
&\varrho\left(\frac{\partial s}{\partial\varrho}\underbrace{\varrho_{,t}}+\frac{\partial s}{\partial\varrho_{,i}}\underbrace{\varrho_{,it}}+\frac{\partial s}{\partial\varepsilon}\underbrace{\varepsilon_{,t}}+\frac{\partial s}{\partial v_k}\underbrace{v_{k,t}}+\frac{\partial s}{\partial v_{k,j}}\underbrace{v_{k,jt}}\right)+\\
&+\varrho v_i\left(\frac{\partial s}{\partial\varrho}\varrho_{,i}+\frac{\partial s}{\partial\varrho_{,j}}\varrho_{,ji}+\frac{\partial s}{\partial\varepsilon}\varepsilon_{,i}+\frac{\partial s}{\partial v_j}v_{j,i}+\frac{\partial s}{\partial v_{j,k}}v_{j,ki}\right)+\\
&+\frac{\partial J_i}{\partial\varrho}\varrho_{,i}+\frac{\partial J_i}{\partial\varrho_{,j}}\varrho_{,ji}+\frac{\partial J_i}{\partial\varepsilon}\varepsilon_{,i}+\frac{\partial J_i}{\partial v_j}v_{j,i}+\frac{\partial J_i}{\partial v_{j,k}}v_{j,ki}\geq 0.
\end{aligned}
\tag{13}
$$

In (13) we did not use for the entropy flux $\mathbf{J}$ the classical assumption of rational thermodynamics $\mathbf{J}=\frac{\mathbf{q}}{\vartheta}$, but we left $\mathbf{J}$ in the undetermined form. In this way, the technique can be applied to a larger class of materials. Moreover, in order to illustrate better our procedure, we have underlined the terms which have to be substituted by using the equations (8)-(12). Such a substitution requires cumbersome calculations and leads to a very long entropy inequality, which is not the case to report here.

Let us limit ourselves to notice that, by imposing that such an inequality is satisfied for arbitrary thermodynamic processes, we can get the following thermodynamic restrictions

$$\frac{\partial s}{\partial v_{i,l}}\frac{\partial T_{ij}}{\partial\varrho_{,k}}=0, \tag{14}$$

$$\frac{\partial s}{\partial v_{i,l}}\frac{\partial T_{ij}}{\partial\varepsilon}=0, \tag{15}$$

$$\frac{\partial s}{\partial v_{i,l}}\frac{\partial T_{ij}}{\partial v_{k,t}} = 0, \tag{16}$$

$$\begin{aligned}
&-\varrho\frac{\partial s}{\partial \varrho_{,l}}v_k - \frac{\partial s}{\partial \varepsilon}\frac{\partial h_l}{\partial \varrho_{,k}} + \frac{\partial s}{\partial v_i}\frac{\partial T_{il}}{\partial \varrho_{,k}} - \frac{1}{\varrho}\varrho_{,j}\frac{\partial s}{\partial v_{i,j}}\frac{\partial T_{il}}{\partial \varrho_{,k}} + \\
&+\frac{\partial s}{\partial v_{i,l}}\frac{\partial^2 T_{ij}}{\partial \varrho \partial \varrho_{,k}}\varrho_{,j} + \frac{\partial s}{\partial v_{i,l}}\frac{\partial T_{ik}}{\partial \varrho} + \frac{\partial s}{\partial v_{i,j}}\frac{\partial^2 T_{il}}{\partial \varrho_{,k}\partial \varrho}\varrho_{,j} + \\
&+\frac{\partial s}{\partial v_{i,j}}\frac{\partial^2 T_{il}}{\partial \varrho_{,k}\partial v_t}v_{j,t} + \frac{\partial s}{\partial v_{i,l}}\frac{\partial^2 T_{ij}}{\partial v_s \partial \varrho_{,k}}v_{s,j} + \varrho v_l \frac{\partial s}{\partial \varrho_{,k}} + \frac{\partial J_l}{\partial \varrho_{,k}} = 0,
\end{aligned} \tag{17}$$

$$\begin{aligned}
&-\frac{\partial s}{\partial \varepsilon}\frac{\partial h_j}{\partial \varepsilon} + \frac{\partial s}{\partial v_i}\frac{\partial T_{ij}}{\partial \varepsilon} - \frac{1}{\varrho}\varrho_{,l}\frac{\partial s}{\partial v_{i,l}}\frac{\partial T_{ij}}{\partial \varepsilon} + \frac{\partial s}{\partial v_{i,j}}\frac{\partial^2 T_{il}}{\partial \varrho \partial \varepsilon}\varrho_{,l} + \frac{\partial s}{\partial v_{i,l}}\frac{\partial^2 T_{ij}}{\partial \varepsilon \partial \varrho}\varrho_{,l} + \\
&+\frac{\partial s}{\partial v_{i,l}}\frac{\partial^2 T_{ij}}{\partial \varepsilon \partial v_t}v_{t,l} + \frac{\partial s}{\partial v_{i,j}}\frac{\partial^2 T_{il}}{\partial v_s \partial \varepsilon}v_{s,l} + \varrho v_j \frac{\partial s}{\partial \varepsilon} + \frac{\partial J_j}{\partial \varepsilon} = 0,
\end{aligned} \tag{18}$$

$$\begin{aligned}
&-\varrho^2\frac{\partial s}{\partial \varrho_{,l}}\delta_{mk} - \frac{\partial s}{\partial \varepsilon}\frac{\partial h_l}{\partial v_{m,k}} + \frac{\partial s}{\partial v_i}\frac{\partial T_{il}}{\partial v_{m,k}} - \frac{1}{\varrho}\varrho_{,j}\frac{\partial s}{\partial v_{i,j}}\frac{\partial T_{il}}{\partial v_{m,k}} - \varrho v_k \frac{\partial s}{\partial v_{m,l}} + \\
&+\frac{\partial s}{\partial v_{i,l}}\frac{\partial^2 T_{ij}}{\partial \varrho \partial v_{m,k}}\varrho_{,j} + \frac{\partial s}{\partial v_{i,l}}\frac{\partial^2 T_{ij}}{\partial v_s \partial v_{m,k}}v_{s,j} + \frac{\partial s}{\partial v_{i,l}}\frac{\partial T_{ik}}{\partial v_m} + \\
&+\frac{\partial s}{\partial v_{i,j}}\frac{\partial^2 T_{il}}{\partial v_{m,k}\partial \varrho}\varrho_{,j} + \frac{\partial s}{\partial v_{i,j}}\frac{\partial^2 T_{il}}{\partial v_{m,k}\partial v_h}v_{h,j} + \varrho v_l \frac{\partial s}{\partial v_{m,k}} + \frac{\partial J_l}{\partial v_{m,k}} = 0.
\end{aligned} \tag{19}$$

Let us observe that the restrictions above allow the presence of the first density gradient into the whole set of constitutive equations. Hence, second grade fluids are capable to respect the constraints following by second law.

3 Equivalence with the extended Liu procedure

As we noticed in Section 1, the classical Liu procedure can be extended by adding to the entropy inequality the balances of the basic variables and of their gradients times suitable Lagrange multipliers [12]. Hence, the Liu inequality can be written as

$$\begin{aligned}
\varrho s_{,t} + \varrho v_i s_{,i} + J_{i,i} - \\
-\lambda^{\varrho}\{Mass\text{Balance}\} - \\
-\lambda^{\varrho}_i\{Extended\text{MassBalance}\} - \\
-\lambda^{\varepsilon}\{Internal\text{EnergyBalance}\} - \\
-\lambda^{\text{v}}_i\{Linear\text{MomentumBalance}\} - \\
-\lambda^{\text{v}}_{il}\{Extended\text{LinearMomentumBalance}\} \geq 0,
\end{aligned} \tag{20}$$

where the quantities multiplying the balances are just the Lagrange multipliers mentioned above. Also in this case, if one imposes that (20) is fulfilled for an arbitrary thermodynamic process, the following thermodynamic restrictions ensue [10, 11]

$$\lambda^{\varrho} = \varrho \frac{\partial s}{\partial \varrho}, \tag{21}$$

$$\lambda_i^{\varrho} = \varrho \frac{\partial s}{\partial \varrho_{,i}}, \tag{22}$$

$$\lambda^{\varepsilon} = \frac{\partial s}{\partial \varepsilon}, \tag{23}$$

$$\lambda_i^{\mathrm{v}} = \frac{\partial s}{\partial v_i} - \frac{1}{\varrho}\lambda_{il}^{\mathrm{v}} \varrho_{,l}, \tag{24}$$

$$\lambda_{il}^{\mathrm{v}} = \frac{\partial s}{\partial v_{i,l}}, \tag{25}$$

$$\lambda_{li}^{\mathrm{v}} \frac{\partial T_{lj}}{\partial \varrho_{,k}} = 0, \tag{26}$$

$$\lambda_{kl}^{\mathrm{v}} \frac{\partial T_{ki}}{\partial \varepsilon} = 0, \tag{27}$$

$$\lambda_{zi}^{\mathrm{v}} \frac{\partial T_{zj}}{\partial v_{l,k}} = 0, \tag{28}$$

$$\begin{aligned} &\varrho v_l \frac{\partial s}{\partial \varrho_{,k}} + \frac{\partial J_l}{\partial \varrho_{,k}} - \lambda^{\varepsilon} \frac{\partial h_l}{\partial \varrho_{,k}} + \lambda_i^{\mathrm{v}} \frac{\partial T_{il}}{\partial \varrho_{,k}} + \lambda_{il}^{\mathrm{v}} \frac{\partial^2 T_{ij}}{\partial \varrho \partial \varrho_{,k}} \varrho_{,j} - \lambda_l^{\varrho} v_k + \\ &-\lambda_{ij}^{\mathrm{v}} \frac{\partial^2 T_{il}}{\partial \varrho_{,k} \partial \varrho} \varrho_{,j} + \lambda_{il}^{\mathrm{v}} \frac{\partial T_{ik}}{\partial \varrho} + \lambda_{ij}^{\mathrm{v}} \frac{\partial^2 T_{il}}{\partial \varrho_{,k} \partial v_z} v_{z,j} + \lambda_{il}^{\mathrm{v}} \frac{\partial^2 T_{ij}}{\partial v_s \partial \varrho_{,k}} v_{s,j} = 0, \end{aligned} \tag{29}$$

$$\begin{aligned} &\frac{\partial J_j}{\partial \varepsilon} - \lambda^{\varepsilon} \frac{\partial h_j}{\partial \varepsilon} + \lambda_i^{\mathrm{v}} \frac{\partial T_{ij}}{\partial \varepsilon} + \lambda_{ij}^{\mathrm{v}} \frac{\partial^2 T_{il}}{\partial \varrho \partial \varepsilon} \varrho_{,l} + \lambda_{il}^{\mathrm{v}} \frac{\partial^2 T_{ij}}{\partial \varepsilon \partial \varrho} \varrho_{,l} + \\ &+\lambda_{il}^{\mathrm{v}} \frac{\partial^2 T_{ij}}{\partial \varepsilon \partial v_z} v_{z,l} + \lambda_{ij}^{\mathrm{v}} \frac{\partial^2 T_{il}}{\partial v_s \partial \varepsilon} v_{s,l} = 0, \end{aligned} \tag{30}$$

$$\begin{aligned} &\varrho v_l \frac{\partial s}{\partial v_{z,k}} + \frac{\partial J_l}{\partial v_{z,k}} - \varrho \lambda_l^{\varrho} \delta_{zk} - \lambda^{\varepsilon} \frac{\partial h_l}{\partial v_{z,k}} + \lambda_i^{\mathrm{v}} \frac{\partial T_{il}}{\partial v_{z,k}} - \\ &-\varrho \lambda_{zl}^{\mathrm{v}} v_k + \lambda_{il}^{\mathrm{v}} \frac{\partial^2 T_{ij}}{\partial \varrho \partial v_{z,k}} \varrho_{,j} + \lambda_{il}^{\mathrm{v}} \frac{\partial^2 T_{ij}}{\partial v_s \partial v_{z,k}} v_{s,j} + \lambda_{il}^{\mathrm{v}} \frac{\partial T_{ik}}{\partial v_z} + \\ &+\lambda_{ij}^{\mathrm{v}} \frac{\partial^2 T_{il}}{\partial v_{z,k} \partial \varrho} \varrho_{,j} + \lambda_{ij}^{\mathrm{v}} \frac{\partial^2 T_{il}}{\partial v_{z,k} \partial v_s} v_{s,j} = 0. \end{aligned} \tag{31}$$

Let us prove now that the restrictions (21)-(31) hold if, and only if, the restrictions (14)-(19) hold too. To this end, let us suppose that equations (21)-(31) are fulfilled for each thermodynamic process. Then, if we substitute (21)-(25) into (26)-(31) we obtain (14)-(19).

Conversely, if (14)-(19) hold, then (26)-(31) are recovered by defining the Lagrange multipliers according to (21)-(25).

In such a way, we have proved that for viscous second grade fluids the extended Coleman-Noll and Liu procedures are equivalent since, if the state space (7) is chosen and if the evolution of the state variables is governed by the system (8)-(12), then they lead to the same set of thermodynamic restrictions.

4 Conclusions

In this paper we have developed a nonlocal model of viscous fluids. Its thermodynamic compatibility has been investigated by applying an extended Coleman-Noll procedure for the exploitation of second law. We have shown that the same results can be carried out through an extended Liu procedure. Both the techniques show that second grade Korteweg type fluids are fully compatible with the restrictions placed by second law of thermodynamics. Such a compatibility does not emerge if one applies the classical Coleman-Noll or Liu procedures.

The extended techniques considered here can lead to new results in several situations in which nonlocal potentials seem to be necessary. This is the case of liquid crystals [14], damage theory [15, 16] and diffusive-hyperbolic heat conduction [17, 18].

It is worth observing that the equivalence proved in Section 3 implies the equivalence of the classical Coleman-Noll and Liu procedures too.

This represents an extension to a larger class of nonlocal materials of an earlier result proved in [13] on the example of rigid heat conductors. The problem of the equivalence of the procedures for general systems whose constitutive equations contain time and space derivatives of the basic variables up to an arbitrary order will be considered in a forthcoming paper.

Acknowledgements

Vita Triani thanks for financial support the *Regione Basilicata* POR 2000-2006, Research Project *Modelli matematici in termodinamica di non equilibrio.*

References

[1] D. J. Korteweg. *Sur la forme qui prennent les equations du mouvement des fluids si l'on tient compte des forces capillaires par des variations de densité.* Arch. Néer. Sci. Exactes, Ser II., **6** (1901), 1-24.

[2] E. C. Aifantis, J. Serrin. *The mechanical theory of fluid interfaces and Maxwell's rule.* J. Colloid Interface Sci., **96** (1983), 517-529.

[3] E. C. Aifantis, J. Serrin. *Equilibrium solutions in the mechanical theory of fluid microstructures.* J. Colloid Interface Sci., **96** (1983), 530-547.

[4] M. Slemrod. *Admissibility criteria for propagating phase boundaries in a van der Waals fluid.* Arch. Rat. Mech. Anal., **81** (1983), 301-315.

[5] M. Slemrod. *Dynamic phase transitions in a van der Waals fluid.* J. Differential Equations, **52** (1984), 1-23.

[6] J. E. Dunn, J. Serrin. *On the thermomechanics of the interstitial working.* Arch. Rat. Mech. Anal., **88** (1985), 95-133.

[7] B. D. Coleman, D. R. Owen. *A mathematical foundation for thermodynamics.* Arch. Rat. Mech. Anal., **54** (1974), 1-104.

[8] B. D. Coleman, W. Noll. *The thermodynamics of elastic materials with heat conduction and viscosity.* Arch. Rat. Mech. Anal., **13** (1963), 167-178.

[9] I-Shi Liu. *Method of Lagrange multipliers for exploitation of the entropy principle.* Arch. Rat. Mech. Anal., **46** (1972), 131-148.

[10] A. Sellitto, V. Triani, V. A. Cimmelli. *Modelling of capillary forces in viscous fluids of grade two.* Proceedings NTFSM-08, 2008, in this volume.

[11] V. A. Cimmelli, A. Sellitto, V. Triani. *A new thermodynamic framework for second grade fluids.* Forthcoming.

[12] V. A. Cimmelli. *An extension of Liu procedure in weakly nonlocal thermodynamics.* J. Math. Phys., **48** (2007), 113510 (13 pages).

[13] V. Triani, C. Papenfuss, V. A. Cimmelli, W. Muschik. *Exploitation of the Second Law: Coleman-Noll and Liu Procedure in Comparison.* J. Non-Equilib. Thermodyn., **33** (2008), 47-60.

[14] C. Papenfuss, S. Forest. *Thermodynamical framework for higher grade material theories with internal variables or additional degrees of freedom.* J. Non-Equilib. Thermodyn., **31** (2006), 319-353.

[15] Z. P. Bazant, G. Pijaudier-Cabot. *Nonlocal damage theory.* J. Eng. Mech. ASCE, **113** (1987), 1512-1533.

[16] P. Ván, B. Vásárhelyi. *Second Law of thermodynamics and the failure of rock materials.* In: Rock Mechanics in the National Interest, Vol.1, J. P. Tinucci, D. Elsworth and K. A. Heasley Eds, Balkema Publishers, Lisse, 2001, 767-773.

[17] V. A. Cimmelli, W. Kosiński. *Nonequilibrium semi-empirical temperature in materials with thermal relaxation.* Arch. Mech., **43** (1991), 753-767.

[18] V. A. Cimmelli, P. Ván, *The effects of nonlocality on the evolution of higher order fluxes in nonequilibrium thermodynamics.* J. Math. Phys. **46** (2005), 112901 (15 pages).

Longtime Behaviour of the Solutions of a Nonlinear Parabolic Reaction-Diffusion Equation under Robin Boundary Data

S. Rionero[*,a], *M. Vitiello*[†,b]

[a]Dipartimento di Matematica ed Applicazioni *R. Caccioppoli*,
Università degli Studi di Napoli *Federico II*

[b]Dipartimento di Matematica,
Politecnico di Bari,
II Facoltà di Ingegneria, Taranto.

Abstract

Under Robin boundary data, the nonlinear partial differential equation

$$u_t = \Delta F(\mathbf{x}, u) + f(\mathbf{x}, u, \nabla u) \qquad (\mathbf{x}, t) \in \Omega \times R^+$$

is studied. Existence and longtime behaviour of the solutions are established.

1 Introduction

Let $\Omega \in R^3$ be a bounded domain. We consider the nonlinear reaction diffusion equation of type

$$u_t = \Delta F(\mathbf{x}, u) + f(\mathbf{x}, u, \nabla u) \qquad (\mathbf{x}, t) \in \Omega \times R^+. \tag{1}$$

The (1) is a very popular model with a great variety of applications, ranging from physics to biology [1]-[6]. Recentely in [9] the existence of the steady states of (1) and their stability have been studied.
In the present paper we reconsider the problem and look for conditions on F and f under which each solution tends towards a steady state, when $t \to \infty$, or is definitively bounded.
The paper is structured as follows. Section 2 deals with some preliminaries followed by an existence result for problem (1) (Section 3). In Section 4 we prove the existence of the global attractor. The blow up in finite time of the solutions is studied in Section 5.

*Email: rionero@unina.it
†Email: vitiello@poliba.it

2 Statement of the problem

Let $\Omega \in R^3$ be an open bounded set with $C^{2+p}(p>0)$ boundary $\partial\Omega$, having the interior cone property. We consider the initial boundary value problem

$$\begin{cases} u_t = \Delta F(\mathbf{x}, u) + f(\mathbf{x}, u, \nabla u) & (\mathbf{x}, t) \in \Omega \times R^+, \\ u(\mathbf{x}, 0) = u_0(\mathbf{x}) & \mathbf{x} \in \Omega, \\ \beta u(\mathbf{x}, t) + (1-\beta)\nabla u(\mathbf{x}, t) \cdot \mathbf{n} = 0 & (\mathbf{x}, t) \in \partial\Omega \times R^+, \end{cases} \tag{2}$$

where $\mathbf{n}$ denotes the outward unit normal vector to $\partial\Omega$, β is a constant $\in [0,1]$, $F \in C^2(R \times \Omega)$, $f \in C(\Omega \times R \times R^n)$ and $u_0 \in C(\overline{\Omega})$.
The associated steady boundary value problem is given by

$$\begin{cases} \Delta F(\mathbf{x}, U) + f(\mathbf{x}, U, \nabla U) = 0 & \mathbf{x} \in \Omega, \\ \beta U + (1-\beta)\nabla U \cdot \mathbf{n} = 0 & \mathbf{x} \in \partial\Omega. \end{cases} \tag{3}$$

We assume that exists a positive constant m such that

$$\frac{\partial F(\mathbf{x}, U)}{\partial u} \geq 2m \qquad a.e. \qquad in \ \ \Omega. \tag{4}$$

Therefore there exists a positive constant ϵ such that [7]-[9]

$$|v| < \epsilon \Rightarrow \frac{\partial F(\mathbf{x}, U+v)}{\partial u} \geq m \quad a.e. \ \ in \ \ \Omega.$$

Moreover, denoting by $\|\cdot\|$ the $L^2(\Omega)$-norm, as concerns f, we assume that exist two positive constants μ and k such that [6]-[7]

$$\left\|f^{n+1}\right\| \leq \mu \left\|u^{n+1}\right\|^{2k+1}. \tag{5}$$

In the sequel we will denote by $W^{1,2}(\Omega, \partial\Omega, \beta)$ the subspace of $W^{1,2}(\Omega)$ such that

$$\Psi \in W^{1,2}(\Omega, \partial\Omega, \beta) \Rightarrow \left\{\Psi \in W^{1,2}(\Omega) \cap W^{1,2}(\partial\Omega); \beta\Psi + (1-\beta)\frac{d\Psi}{d\mathbf{n}} = 0 \ \ on \ \ \partial\Omega \times R^+\right\}.$$

We recall that in $W^{1,2}(\Omega, \partial\Omega, \beta)$ the following inequality holds [10]-[11]

$$\|\nabla\Psi\|^2 + \frac{\beta}{1-\beta}\int_{\partial\Omega} \Psi^2 d\Sigma \geq \overline{\lambda}\,\|\Psi\|^2. \tag{6}$$

As it is well known, $\overline{\lambda}$ is the lowest eigenvalue in $W^{1,2}(\Omega, \partial\Omega, \beta)$ of

$$\Delta\Psi + \lambda\Psi = 0. \tag{7}$$

3 Existence result

This section is devoted to the existence of the solutions of (2). The following theorem holds

Theorem 1. *Suppose that (4) and (5) hold. Then there exists a solution u to problem (2), satisfying the following conditions*

- $u \in L^2(\Omega)$.
- *(2)$_1$ holds in* $L^2(\Omega)$.

Proof. First of all, we consider the orthonormal basis $\{w_k : k \in N\}$ of $\mathrm{L}^2(\Omega)$, where $w_k \in H^1(\Omega)$ is an eigenfunction of the following spectral problem (7), i.e.

$$\begin{cases} -\Delta w_k = \lambda_k w_k & \mathbf{x} \in \Omega, \\ \beta w_k + (1-\beta)\dfrac{dw_k}{d\mathbf{n}} = 0 & \mathbf{x} \in \partial\Omega. \end{cases} \tag{8}$$

We also normalize w_k such that $||w_k|| = 1$.
Now we employ the Faedo-Galerkin method: for each integer n we will look for an approximate solution u_n of the form

$$u_n(t) = \sum_{k=1}^{n} g_{kn}(t) w_k(x) \tag{9}$$

satisfying

$$\begin{cases} (u_n', w_k) = (\Delta F(\mathbf{x}, u_n), w_k) + (f(\mathbf{x}, u_n, \nabla u_n), w_k) & k = 1, 2, ..., n \\ (u_n(0), w_k) = \alpha_{kn} \end{cases} \tag{10}$$

where α_{kn} are suitable constants such that

$$\lim_{n \to \infty} \sum_{k=1}^{n} \alpha_{kn} w_k = u_0.$$

Then (10) is reduced to the following initial value problem for the system of nonlinear ordinary differential equations (with respect to g_{kn}) given by

$$\begin{cases} \dfrac{dg_{kn}}{dt} = G_k(g) + f_k(g) & k = 1, 2, ...n, \\ g_{kn}(0) = \alpha_{kn}, \end{cases} \tag{11}$$

where

$$G_k(g) = \int_\Omega \Delta F(\mathbf{x}, \sum_{k=1}^{n} g_{kn} w_k) w_k d\Omega, \quad \textit{and} \quad f_k(g) = \int_\Omega f(\mathbf{x}, \sum_{k=1}^{n} g_{kn} w_k, \sum_{k=1}^{n} g_{kn} \nabla w_k) w_k d\Omega.$$

According to the standard existence theory for ordinary differential equations, there exists a local solution in the interval $(0, T_n)$ of (11).
In order to extend the local solution to the interval $(0, T)$ independent of n, the following estimates are needed.
Estimate 1: We multiply the equation $(10)_1$ by $g_{kn}(t)$, adding these relations by $k = 1, 2, ..., n$, we obtain

$$\frac{1}{2}\frac{d}{dt}||u_n||^2 = \int_{\partial\Omega} \nabla F(\mathbf{x}, u_n)\cdot \mathbf{n} u_n d\Sigma - \int_\Omega \frac{\partial F(\mathbf{x}, u_n)}{\partial u}(\nabla u_n)^2 d\Omega + \int_\Omega f(\mathbf{x}, u_n, \nabla u_n) u_n d\Omega.$$

From the boundary condition $(2)_3$, (4) and (5) it turns out that

$$\frac{1}{2}\frac{d}{dt}||u_n||^2 \leq -\frac{\beta m}{1-\beta}\int_{\partial\Omega} u_n^2 d\Sigma - m||\nabla u_n||^2 + \mu||u_n||^2.$$

By virtue of (6) it follows that

$$\frac{1}{2}\frac{d}{dt}||u_n||^2 + \left(m - \frac{\mu}{\overline{\lambda}}\right)\left[\frac{\beta}{1-\beta}\int_{\partial\Omega} u_n^2 d\Sigma + m||\nabla u_n||^2\right] \leq C$$

with C positive constant. Integrating over $(0, t)$ one obtains

$$||u_n||^2 + 2\left(m - \frac{\mu}{\overline{\lambda}}\right)\int_0^t \left[\frac{\beta}{1-\beta}\int_{\partial\Omega} u_n^2 d\Sigma + m||\nabla u_n||^2\right] dt \leq C_T \tag{12}$$

where C_T is a positive constant depending only on $||u_0||$ and T.
Thus, (12) yields that (if $\mu \leq \overline{\lambda} m$)

$$(u_n) \quad \textit{is} \quad \textit{bounded} \quad \textit{in} \quad L^2(\Omega). \tag{13}$$

Estimate 2: Multiplying the equation $(10)_1$ by g'_{kn}, then summing with respect to k, we obtain

$$||u_n'||^2 \leq \int_{\partial\Omega} \frac{\partial F(\mathbf{x}, u_n)}{\partial u}\nabla u_n\cdot\mathbf{n} u_n' d\Sigma - \int_\Omega \frac{\partial F(\mathbf{x}, u_n)}{\partial u}\nabla u_n \nabla u_n' d\Omega + \int_\Omega f(\mathbf{x}, u_n, \nabla u_n)\cdot u_n' d\Omega.$$

Taking into account the boundary condition $(2)_3$ and (4), the Cauchy-Schwarz inequality implies

$$||u_n'||^2 + \frac{m}{2}\frac{d}{dt}\left[\frac{\beta}{1-\beta}\int_{\partial\Omega} u_n^2 d\Sigma + ||\nabla u_n||^2\right] \leq \frac{1}{2}||f(\mathbf{x}, u_n, \nabla u_n)||^2 + \frac{1}{2}||u_n'||^2.$$

From (5) it turns out that

$$\frac{1}{2}||u_n'||^2 + \frac{m}{2}\frac{d}{dt}\left[\frac{\beta}{1-\beta}\int_{\partial\Omega} u_n^2 d\Sigma + ||\nabla u_n||^2\right] \leq \frac{\mu^2}{2}||u_n||^2.$$

Using (13) we obtain that

$$(u_n') \quad \textit{is} \quad \textit{bounded} \quad \textit{in} \quad L^2(\Omega) \tag{14}$$

Limit of the approximate solutions: Estimate (13) and (14) allow us to pass to the limit in each term of (10). For any fixed k, letting $n \to \infty$ in (10), we obtain that the following holds in $L^2(\Omega)$

$$(u', w_k) = (\Delta F(\mathbf{x}, u), w_k) + (f, w_k).$$

4 Global attractor

In this section we study the existence of a global attractor.

Theorem 2. *Let (4) and (5) hold. For $u_0 \in L^2(\Omega)$ the problem (2) has a global attractor that is compact and connected.*

Proof. To prove that problem (2) admits a compact connected global attractor, the first step is to prove that exists an absorbing set in $L^2(\Omega)$ and then the existence of an absorbing set in $H^1(\Omega)$.
Absorbing set in $L^2(\Omega)$.
Multiplying the equation $(2)_1$ by u, then integrating over Ω, yields

$$\frac{1}{2}\frac{d}{dt}||u||^2 = \int_{\partial\Omega} \frac{\partial F(\mathbf{x},u)}{\partial u} u \nabla u \cdot \mathbf{n} d\Sigma - \int_{\Omega} \frac{\partial F(\mathbf{x},u)}{\partial u} (\nabla u)^2 d\Omega + \int_{\Omega} f(\mathbf{x}, u, \nabla u) u d\Omega. \quad (15)$$

Therefore, taking into account the boundary conditions $(2)_3$ and (4), it follows that

$$\frac{1}{2}\frac{d}{dt}||u||^2 \leq -\frac{m\beta}{1-\beta}\int_{\partial\Omega} u^2 d\Sigma - m||\nabla u||^2 + \int_{\Omega} f(\mathbf{x}, u, \nabla u) u d\Omega. \quad (16)$$

The Holder inequality and (6) give

$$\frac{1}{2}\frac{d}{dt}||u||^2 + (m\overline{\lambda} - \mu)||u||^2 \leq C_2 \quad (17)$$

with C_2 positive constant.
Let

$$E(t) = \frac{1}{2}\int_{\Omega} u^2 d\Omega \quad (18)$$

be a certain energy function, (17) implies

$$\frac{d}{dt}E(t) + 2C_1 E(t) \leq C_2$$

with C_1, C_2 positive constants indipendent of initial data u_0.
By solving this differential inequality, with $\mu < \overline{\lambda} m$, we see that

$$E(t) \leq E(0)e^{-2C_1 t} + \frac{C_2}{2C_1}(1 - e^{-2C_1 t}). \quad (19)$$

Thus

$$\lim_{t\to\infty} sup E(t) \leq \rho_0^2 \quad (20)$$

with $\rho_0^2 = \dfrac{C_2}{2C_1}$.
There exists an absorbing set B_0 in $L^2(\Omega)$, namely, any ball of $L^2(\Omega)$ centered at O of radius $\rho_0' > \rho_0$. If B is a bounded set of $L^2(\Omega)$, included in a ball $B(O, R)$ of $L^2(\Omega)$, centered at O of radius R, then

$$u(t) \in B(O, \rho_0') \qquad \forall t \geq t_0(B, \rho_0')$$

with

$$t_0 = \frac{1}{2C_1} log \frac{R^2}{(\rho_0')^2 - \rho_0^2}.$$

We also note that the ball $B(0, \rho_0')$, for $\rho_0' > \rho_0$ and even $\rho_0' = \rho_0$ are positively invariant. In fact, from (19), if $E(0) < \rho_0'^2$, we infer that

$$E(t) < \rho_0'^2 e^{-2C_1 t} + \rho_0^2(1 - e^{-2C_1 t}) =$$

$$= (\rho_0'^2 - \rho_0^2)e^{-2C_1 t} + \rho_0^2 \leq \rho_0'^2.$$

Now we observe that from (17) it turns out that

$$\frac{1}{2}\frac{d}{dt}||u||^2 + \left(m - \frac{\mu}{\lambda}\right)\left[\frac{\beta}{1-\beta}\int_{\partial\Omega} u^2 d\Sigma + ||\nabla u||^2\right] \leq C_2.$$

Integrating from t to $t + \delta$, it follows that

$$\left(m - \frac{\mu}{\lambda}\right)\int_t^{t+\delta}\left[\frac{\beta}{1-\beta}\int_{\partial\Omega} u^2 d\Sigma + ||\nabla u||^2\right] d\tau \leq C_2\delta + E(t).$$

Thanks to (20) we obtain

$$\lim_{t\to\infty} sup \left\{\int_t^{t+\delta}\left[\frac{\beta}{1-\beta}\int_{\partial\Omega} u^2 d\Sigma + ||\nabla u||^2\right] d\tau\right\} \leq C_2\delta + \rho_0^2. \tag{21}$$

If $u_0 \in B(O, \rho_0')$ and $t > t_0$, (21) becomes

$$\int_t^{t+\delta}\left[\frac{\beta}{1-\beta}\int_{\partial\Omega} u^2 d\Sigma + ||\nabla u||^2\right] d\tau \leq C_2\delta + (\rho_0')^2.$$

Absorbing set in $H^1(\Omega)$

We now prove the existence of an absorbing set in $H^1(\Omega)$ and the uniform compactness. For this purpose we multiply $(2)_1$ by $-\Delta F(\mathbf{x}, u)$. Integrating over Ω and using the divergence theorem we obtain

$$-\int_{\partial\Omega} \frac{\partial F(\mathbf{x}, u)}{\partial u} \nabla u \cdot \mathbf{n} u_t d\Sigma + \int_\Omega \frac{\partial F(\mathbf{x}, u)}{\partial u} \nabla u \nabla u_t d\Omega =$$

$$= -\int_\Omega (\Delta F(\mathbf{x}, u))^2 d\Omega - \int_\Omega f(\mathbf{x}, u, \nabla u) \cdot \Delta F(\mathbf{x}, u) d\Omega.$$

By $(2)_3$ and (4) one obtains

$$\frac{1}{2} m \frac{d}{dt}\left[\frac{\beta}{1-\beta}\int_{\partial\Omega} u^2 d\Sigma + \int_\Omega (\nabla u)^2 d\Omega\right] \leq -||\Delta F||^2 - \int_\Omega f(\mathbf{x}, u, \nabla u) \cdot \Delta F(\mathbf{x}, u) d\Omega. \tag{22}$$

Let us introduce the following Liapunov function

$$\Sigma(t) = \frac{1}{2}\left[\frac{\beta}{1-\beta}\int_{\partial\Omega} u^2 d\Sigma + \int_\Omega (\nabla u)^2 d\Omega\right], \tag{23}$$

from (22) one obtains

$$m\dot{\Sigma}(t) \leq -||\Delta F||^2 - \int_\Omega f \cdot \Delta F d\Omega.$$

The Cauchy-Schwarz inequality, the Poincaré inequality (6) and (5) imply

$$\dot{\Sigma}(t) \leq \frac{\mu^2}{\overline{\lambda} m}\Sigma(t).$$

The application of the uniform Gronwall Lemma, for an arbitrary fixed $\delta > 0$, gives

$$\forall\delta > 0 \quad \Sigma(t+\delta) \leq \frac{k}{\delta} e^{\frac{\mu^2 \delta}{\overline{\lambda} m}}$$

provided

$$\int_t^{t+\delta} \Sigma(\tau) d\tau \leq k \qquad \forall t \geq t^*. \tag{24}$$

We observe that from (21), (24) holds with

$$k = C_2\delta + R^2$$

Hence (21) provides a uniform bound for $||\nabla u(t+\delta)||$ for $0 \leq t \leq \delta$. For our purpose it is simpler and sufficient to set $t^* = t_0$; in this case, the value of k is given by

$$k = \delta C_1 + \rho_0'^2. \tag{25}$$

It follows that the ball of $H^1(\Omega)$ centered at O of radius ρ_1 is absorbing in $H^1(\Omega)$, where $\rho_1 = \frac{k}{\delta} exp(\frac{\mu^2\delta}{\overline{\lambda} m})$. If u_0 belongs to the ball B(O, R) of $\mathrm{L}^2(\Omega)$ centered at O of radius R, then $u(t)$ enters this absorbing set denoted B_1 at time $t \leq t_0 + \delta$, and remains in it for $t \geq t_0 + \delta$.
At the same time, this result provides that any bounded set B of $\mathrm{L}^2(\Omega)$ is included in such a ball B(O, R), and for $u_0 \in B$ and $t \geq t_0 + \delta$, t_0, δ, as above, $u(t)$ belongs to B_1 which is bounded in $H^1(\Omega)$ and relatively compact in $L^2(\Omega)$.

5 Blow Up

Assuming

$$\frac{\partial F(\mathbf{x}, u)}{\partial u} \leq -m^* \qquad m^* = positive \;\; const. \tag{26}$$

then the following theorem holds true

Theorem 3. *Let (26), (5) and*

$$2m^*\overline{\lambda} < 1 \tag{27}$$

hold true. Then u blow up in a finite time.

Proof. In view of (15) and (18), it follows that

$$\frac{1}{2}\frac{d}{dt}E(t) \geq m^* \left[||\nabla u||^2 + \frac{\beta}{1-\beta}\int_{\partial\Omega} u^2 d\Sigma\right] + \int_\Omega f(\mathbf{x}, u\nabla u)ud\Omega.$$

By virtue of the Poincaré inequality (6) and the Cauchy-Schwarz inequality, one obtains

$$\dot{E}(t) \geq (2m^*\overline{\lambda} - 1)E(t) - \frac{1}{2}||f(\mathbf{x}, u\nabla u)||^2.$$

Then, in view of (18) and (5), it turns out that

$$\dot{E}(t) \geq (2m^*\overline{\lambda} - 1)E(t) - \mu 2^{k+1}E^{2k+1}(t). \tag{28}$$

Then, integrating, one obtains

$$E^{2k}(t) \geq \frac{\eta E^{2k}(0)e^{\eta t}}{\eta + \eta_1 E^{2k}(0)e^{\eta t}}$$

with

$$\eta = 2k(2m^*\overline{\lambda} - 1) \quad \mathit{and} \quad \eta_1 = 2^{2k+2}k\mu^2.$$

This implies that when $t \to T$, with

$$T = \frac{1}{\eta}\log\left(\frac{2m^*\overline{\lambda}}{\mu 2^{k+1}E^{2k}(0)}\right), \tag{29}$$

the solution u, in view of (27), blow up at time (29).

Acknowledgements

This work has been performed under the auspices of the G.N.F.M. of I.N.D.A.M. and M.I.U.R. (P.R.I.N.): *Waves and Stability in Continuous Media.*

References

[1] J. N. Flavin and S. Rionero, *Qualitative Estimates for Partial Differential Equation, An Introduction*, CRC Press, Boca Raton, FL, (1995).

[2] J. Smoller, *Shock waves and Reaction-Diffusion Equations*, Springer-Verlag, n.258 of "A Series of Comprehensive Studies in Mathematics", (1983).

[3] J. D. Murray, *Mathematical Biology*, Biomathematics Text, 19, Springer-Verlag, (1989).

[4] Lawrence C. Evans, *Partial differential Equations*, 1992.

[5] F. Capone, S. Rionero and I. Torcicollo, *On the stability of solutions of the remarkable equation* $u_t = \Delta F(x,u) - g(x,u)$, Suppl. Rend. Cir. Mat. Palermo, serie II, n.45, (1996).

[6] S. Rionero, *Asymptotic properties of solutions to nonlinear possibly degenerated parabolic equations in unbounded domains*, Math. and Mech. of Solids,(2003).

[7] S. Rionero, *A rogorous reduction of the L^2-stability of the solutions to a nonlinear binary reaction-diffusion system of P.D.Es to the stability of the solutions to a linear binary system of O.D.Es.*, Journal of math. Analysis and Application 319, 372-392, (2006).

[8] S. Rionero, *On the longtime behaviour of the solutions of nonlinear parabolic equations in unbounded domains*, Proceedings Wascom 2001, World Scientific, Singapore, 447-457, (2002).

[9] S. Rionero, M. Vitiello, *Stability properties of the reaction diffusion equation with Robin boundary conditions*, (to appear on Proceedings of WASCOM 2008).

[10] K. Rektoris, *Varational methods in Mathematics*, Science and Engineering D. Reidd Pubb. Co., (1980).

[11] Senn, *On a nonlinear elliptic eigenvalue problem with Neumann boundary conditions with an application to population dynamics*, Comm. Partial Diff. Eq. 8, n. 11, 1199-1228, (1983).